THE MAKING
OF THE MIND

The Neuroscience
of Human Nature

THE MAKING
OF THE MIND

RONALD T. KELLOGG

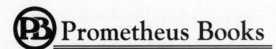

Prometheus Books

59 John Glenn Drive
Amherst, New York 14228–2119

Published 2013 by Prometheus Books

Cover image © 2013 Media Bakery
Cover design by Grace M. Conti-Zilsberger

Library of Congress Cataloging-in-Publication Data

Kellogg, Ronald Thomas.
    The making of the mind : the neuroscience of human nature / by Ronald T. Kellogg.
        pages cm
    Includes bibliographical references and index.
    ISBN 978-1-61614-733-4 (pbk.)
    ISBN 978-1-61614-734-1 (ebook)
    1. Cognitive neuroscience. 2. Neuropsychology. I. Title.

QP360.5.K45 2013
612.8'233—dc23

2013010071

Printed in the United States of America

CONTENTS

Chapter 1

ORIGINS

T he origin of the human mind is as lost as a grain of sand in the Sahara. It is impossible to say precisely when the psychological ingredients of modern human minds made their entry into the world because minds do not leave fossils that can be dated and interpreted by scientists. Today, we can examine the living organ of the mind—the human brain—as it perceives, thinks, remembers, and imagines by means of the recently invented tools of neuroimaging. Ancestral brains, on the other hand, turned to dust long ago, whether their owners once lived thousands, tens of thousands, or hundreds of thousands of years in the past. All that remain now are portions of the skull that once housed a human mind, or the bones of the legs that enabled upright walking, or the hands that enabled the crafting of tools or visual artistry. Although no one knows for certain when the modern human mind first appeared, archeologists point to abundant evidence of its activity dating from forty thousand to ten thousand years ago in the Upper Paleolithic or Late Stone Age. By then, the mind within us was at work in the world creating ancient art of unmistakably human origin.

In the south of France, not far from Cahors, in the Midi-Pyrénées region, is a cave called Pech Merle. My visit there in the summer of 2005 was prompted by the chance to see prehistoric cave paintings created by early modern humans. At first appearances, the cave at Pech Merle was indistinguishable from those limestone caves carved by water in the valley of the Mississippi River back home in St. Louis. The crucial difference appeared only deep in the cave, far from the entrance, when the tour guide shined his flashlight on a wall adorned with hauntingly beautiful images of human hands and spotted horses.

These and many other images in the cave were created by human beings in the Upper Paleolithic era of prehistory. Using carbon 14 dating, archeologists estimate the age of the spotted horse at about 24,640 years old.[1]

Because of the location of the painting in Pech Merle, it can be inferred that the artist or artists had specific memory abilities well known in contemporary cognitive neuroscience. The artist responsible for the spotted horses had to retrieve an image of a horse from long-term memory and maintain it in a temporary store referred to as working memory. By comparing a visual-spatial representation in working memory with the image unfolding on the cave wall, the artist would have been able to create an accurate representation. Because the work was done deep in the cave, the model for the drawing was not present in visual sight. Indeed, the artist had to work from some kind of torch light to even see the creation underway. Conceivably, the artist might have been recollecting a specific horse seen at some point in the past, using a capacity for mental travel back in time, or it may have been a more general memory of the concept rather than a specific instance. Alternatively, perhaps the artist was simply imagining a horse that could have differed from any specific horse ever before encountered. Cognitive anthropologists have in fact wondered whether the images were a product of the hallucinogenic imagery of a shaman. Although it had been thought by biologists that wild, predomestic horses had only coats of black or bay, the spotted pattern found in the images of Pech Merle in fact were painted from memory rather than fantasy. The genes responsible for the spotted-leopard pattern found in some modern horses were also identified in DNA samples of horse remains in Western Europe dated to the Pleistocene, the geological epoch ending ten thousand years ago, just prior to the Holocene epoch of today. The same pattern was not found in Asian samples from the Pleistocene, suggesting that all horse colors observed in Paleolithic art in fact were found in the prehistoric horse populations where the caves were located.[2]

The artwork of the Upper Paleolithic provides vivid evidence of a modern human mind. It indicates that modern humans were living in Europe by forty thousand years ago, although they likely originated earlier in Africa, as will be discussed later in the chapter. The makers of the cave paintings and carved figurines were fellow members of *Homo sapiens* just like us today. Archeologists have discovered thousands of pieces of Paleolithic art rendered on limestone cave walls as well as small objects of stone, bone, antler, and ivory that are associated with the fossils of human remains. Although such artwork has been found in several parts of the world, the most intensively investigated and well-known collection comes from southwestern Europe. In the cave at Lascaux in southern France alone, for example, more than fifteen hundred images adorn the walls.[3] Animals are a frequent subject matter, but images of humans, and anthropomorphic-type images that combine parts of an animal (e.g., antlers) with a human figure, are also common. Abstractions in the form of geometric signs are also observable. Unmistakably, the creators of these works were capable of using visual symbols to represent ideas.

During the Upper Paleolithic, there is clear evidence of an explosion of cultural creativity. Besides the cave paintings and sculptures, archeologists document an increased sophistication in tool making and weaponry, and a wide variety of body adornments. This explosion of Upper Paleolithic culture has been referred to as a "human revolution" or the "dawn of human culture" or "a great leap forward" as a way of highlighting the discontinuity with the artifacts produced by earlier members of the hominid lineage.[4] For example, neither *Homo neanderthalis* nor *Homo erectus* left artifacts comparable in their sophistication, variety, and sheer number as those created by *Homo sapiens*. Scholars disagree as to whether the change to the modern mind was as sudden as implied by the concept of "a great leap forward" or instead reflected a gradual accumulation of changes starting back even further in the past.[5] In either case, by forty thousand years ago humanity had crossed the Rubicon, setting the mind within us apart from all other creatures. How is the human mind so fundamentally distinctive? The aim of this book is to provide a novel answer to this question by drawing on contemporary research in cognitive social neuroscience.

By analyzing what has been learned in recent decades about the neuroscience of our cognitive and social abilities, it is possible to identify the most important distinctive features of the mind within us. Major advances have been made with the invention of tools, such as functional magnetic resonance imaging (fMRI), that allow one to infer neural activity as cognitive functions take place in the living brain. Importantly, the brain of a twenty-first-century human being has the same structures and distinctive cognitive capacities as the brain of the Upper Paleolithic artist. It was this brain—the one we can investigate today with the tools of contemporary neuroscience—that enabled us to invent culture and exploit cultural change over relatively brief periods of historical time. The capacity for logical inference exhibited by the scientists who analyzed the DNA of horses to draw conclusions about the cave paintings of the Upper Paleolithic was already present in the prehistoric originators of the art. In like manner, the capacity for planning that enabled humans in the twentieth century to land on the moon was already manifest in our prehistoric ancestors. As the noted paleontologist Stephen Jay Gould observed, "Cultural change has brought most of us through hunting and gathering, past the explosive new world triggered by agriculture, and into the age of atomic weaponry, air transportation and the electronic revolution." This cultural change was brought about with "the same brain that enabled some of us to paint the caves of Chauvet and the ceiling of the Sistine Chapel."[6]

Evolutionary theorists explain the constancy of the human brain over the past forty thousand years in one of two ways. The classical Darwinian view of gradual change over vast periods of geological time holds that tens of thousands of years is insufficient time for significant biological change to occur. A different view, and the one favored by Gould, is that species typically remain static in form over tens of thousands of years with no gradual change at all. When a new species emerges through natural selection, speciation takes place relatively rapidly—as a saltation rather than a slow gradual accumulation of differences. The abrupt transition associated with the formation of a new species is then followed by a new equilibrium where no change at all is observable for long periods of time. This alternative to Darwinian gradualism is called the punctuated equilibrium model of evolution.[7] In any event, the essential point for the thesis of this book is that we can understand the causes of the "human revolution" during the Upper Paleolithic by investigating the brain and mental capabilities of modern human beings alive today. With the recent advances in the cognitive sciences fueled in part by the invention of technology for imaging the living brain, it is now possible to outline why the mind within us is so distinctive.

The premise of stasis in the fundamental distinctive features of the human does not conflict with the fact that some simple adaptations have occurred over the past ten thousand years. For example, with the invention of agriculture and the domestication of cattle in the final Neolithic phase of the Stone Age (8000–9000 BCE), a dietary adaptation occurred to allow the consumption of milk during adulthood. The hunter-gatherers of prehistory needed the lactase enzyme required to digest milk early in life while nursing, but then it was turned off during maturation. A mutation in the gene responsible for turning off the production of lactase allowed even adults to consume milk, an adaptation that eventually spread widely among Europeans.[8] The kind of adaptations that resulted in the modern human brain—something akin to adding wings with which to fly—were far too complex to occur rapidly in thousands of years. An example is the human capacity to innovate and plan new ways of living, as was required for the invention of agriculture in the first place.

THE ENSEMBLE HYPOTHESIS

One straightforward and common answer to the mystery of "human revolution" is that the brain expanded or reorganized symbolic thought. Symbols are certainly evident in the cave paintings in the form of representational art. A visual symbolic representation is used to refer to a horse, a bull, a deer, a bison, or another human being. A capacity for symbolic thought could have facilitated the ability to think about abstract concepts that are not tied directly to perceptual experience. Closely related to this hypothesis is that the brain expanded or reorganized to support language, where words are used as symbolic representations, and the use of visual-spatial symbols. If either of these hypotheses is correct, then there should be evidence from neuroscience of an obvious change in the brain from nonhuman to human species. Or, coming at the question from the level of the genes responsible for the development of the brain, there ought to be a mutation in a gene or set of genes that accounts for the distinctive feature of human brain maturation related to symbol use or language.

The thesis of the present book adopts a related but more complex point of view. The argument here is that there was not a single major addition to or reorganization of the brain and cognitive functioning in the advent of modern human beings. Instead, the findings of cognitive social neuroscience document the addition of five parts that together comprise the modern mental ensemble. Each part reflected a reorganization of the brain of our immediate hominid ancestors—in some cases this reorganization could have been relatively minor, such as an advance in the executive functioning of working memory; in other cases, the reorganization probably was more significant, such as in the step to using abstract symbols in thought rather than concrete perceptual representations. The central point is that changes occurred on multiple fronts, from small to moderate. Looking at each part in isolation, it might have been possible to discern with ease the continuity with (and specific nature of) the change from our immediate ancestor, had the evidence not long ago turned to dust. But small to modest changes took place on several fronts. This certainly meant that a greatly expanded brain needed to accommodate the sum of the changes. Besides a change in brain quantity, a profoundly important change in quality also occurred because each part interacted with one or more other parts. The parts must be considered together, with each understood only in relation to the whole. The ensemble hypothesis contends that the interaction of the parts, not just the increase in brain size, yielded the dramatically different mind within a modern human.

Here human symbolic thought and language are viewed as so closely interwoven as to constitute just one of the five key parts. Contemporary research in the neurosciences has extensively documented four others that are crucial for the ensemble. One is an advanced working memory, particularly the executive functions that enable planning and self-regulation. Another is an advanced social intelligence that enables humans to collaborate, empathize, and transmit culture from one generation to the next through imitation and other means of social learning. The next part involves a modification of language for the purpose of silent thought rather than vocal communication. As language became interiorized as inner speech, it combined with a capacity to make causal inferences. This part is called *the interpreter* because it explains the contents of our conscious experiences. Lastly, long-term memory in humans underwent a specialization in the ability to recollect specific episodes from our past experience. This episodic form of long-term memory allows one to recall what, when, and where a specific event occurred. Further, it allows one to imagine an event occurring in the future using the same mental capacity. For this reason, the fifth and final part of the ensemble is referred to as *mental time travel*. The modern human mind can roam into the past of our autobiographical experience or venture forward in the imaginary future with equal ease.

When considered in isolation, it should not be surprising to see substantial evidence of the precursors of these five distinctive parts in species with which we share a genetic history. Because the modern chimpanzee is the only living species with a relatively high degree of similarity in its genome to *Homo sapiens*, neuroscientists are especially interested in comparisons between the two species. As will be seen later in the chapter, such comparisons are of limited value because human beings are not descended from modern chimpanzees. *Homo erectus* is hypothesized to either be a direct ancestor of modern human beings or a close side branch of the human lineage; it is in any case extinct and known only from the evidence of paleontology and archeology. *Homo neanderthalis* is another extinct hominid, but one thought to have coexisted with modern human beings recently enough to have left behind DNA samples for geneticists to study. Neither DNA nor fossilized skulls tell us how the brain was organized, however. It is impossible to compare, then, the brain of an immediate ancestor or close hominid relative with the brain of a modern human being.

So, for better or worse, comparative neuroscience does what it can and contrasts the brain and cognitive abilities of chimpanzees with those of human beings. If the ensemble hypothesis is correct, then it should be expected that, say, working memory ought to show continuity between a chimpanzee and a human being. Although the executive functions of human working memory are more advanced than those of a chimpanzee, the differences ought to be a matter of degree rather than of kind. Conceivably, the summation of all five parts yielded a human mind that *is* distinct in kind from a chimpanzee's mind, but that is debatable. According to the ensemble hypothesis, however, the summation of five parts is only the beginning of what sets the mind within us apart. It is the interaction of two or more parts that yields a qualitatively different kind of mind.

The 'human revolution" came as a result of a nonlinear discontinuity with ancestral species, according to the ensemble hypothesis. Advanced working memory is only one of the five parts in the ensemble. The other parts each added their own distinctive features, and critically, they interacted with each to yield a new *kind* of mind in *Homo sapiens*. Thus, the human mind is more than the summation of an advanced working memory plus an advanced social intelligence. The interaction of these two parts alters mental functioning beyond their linear combination. As language, the interpreter, and mental time travel add to the ensemble, the human mind diverges sharply and discontinuously from all nonhuman forms, despite the fact that each individual part shares similarities with that of the nonhuman counterpart. Even though each of the brain reorganizations underlying the components of the ensemble could have been relatively minor compared with an immediate ancestor, the whole of their ultimate effect would be unpredictable from looking at each of the parts. Thus, when considered as an ensemble, the human mind inhabits a mental world on earth with one and only one member.

What exactly is meant by *nonlinear discontinuity*? The critical concept at the heart of the ensemble hypothesis is a statistical interaction of two or more independent factors that have an effect on brain or cognitive complexity. Advanced working memory increases cognitive complexity. So, too, does advanced social intelligence. Both of these operating at once in the modern human brain could increase cognitive complexity in an additive way—the net result may be simply a linear combination of each factor taken separately. Yet, according to the ensemble hypothesis, there is a two-way interaction of factors that yields a nonlinear combination of their independent effects. The power of an advanced social intelligence to boost cognitive complexity is enhanced by the presence of an advanced working memory. Similarly, the power of symbolic thought and language to boost cognitive complexity is impressive on its own merits. Even so, when combined with an advanced social intelligence at the same time, the result is a nonlinear discontinuity from nonhuman minds that possess less powerful social intelligence and modest to minimal symbolic thought and language capabilities. Analogous cases can be made from pairwise interactions of all five parts of the ensemble hypothesized here.

A simple visual metaphor for grasping the idea behind the ensemble hypothesis is to think of five jigsaw puzzle pieces. When looking at any one piece, one can readily see the similarity between it and another piece, a close relative that is slightly or even moderately changed from it. Yet, when the five pieces are assembled, they together portray a picture that is qualitatively different from the appearance of any single piece. In other words, while a family resemblance can be detected with ease at the level of an individual piece, a qualitatively new entity emerges from the ensemble of parts.

An alternative metaphor is to think of the ensemble hypothesis in terms of emergent properties in the physical science of chemistry. Neither the properties of hydrogen (H) nor the properties of oxygen (O) reveal the emergent properties of water when the constituents are appropriately bonded as H_2O. The wetness experienced in the warm waters of a bath or the cool refreshment experienced in drinking water to quench a thirst emerges from the ensemble. Neither experience with hydrogen nor experience with oxygen prepares one for the experience of the warm wetness and cool refreshment of H_2O. The whole is unrecognizable from the parts.

In the remainder of the chapter, foundational facts about the genome and brain of *Homo sapiens* are presented as background for what follows in the book. The thesis that five distinctive features of human cognition emerged in the modern human brain implies that the advent of our species was a highly unlikely event that occurred once and only once. Archeological evidence on this point has long been controversial, with some scholars arguing for a multiregional model of human origins in which the transition to modernity occurred independently in different regions of the world. Recent evidence from human genetics appears to have settled the question in favor of a single origin. It is also important to understand the rationale of comparative neuroscience as an investigative tool into human origins. Comparisons are often made between human brains and behaviors and those of other primates, particularly the common chimpanzee and bonobo. Macaque monkeys, and even rats, commonly provide animal models useful in the neuroscience of human cognition. The basis for, and limitations of, these comparisons can be found in similarities among species with respect to genomes and brain architectures.

INSIGHTS FROM THE GENOME

Recent advances in human genetics have provided scientists with a new tool for understanding the origins of modern humans. To review some of the basics, with the exception of the egg and sperm cells, the nucleus of every cell in the human body contains twenty-three pairs of chromosomes. The final pair is significant because it distinguishes males and females. This twenty-third pair consists of an X-chromosome and a Y-chromosome in males, and two X-chromosomes in females. The egg and sperm cells possess twenty-three single chromosomes rather than pairs that then combine during sexual intercourse and fertilization. Egg cells contain an X-chromosome whereas sperm cells contain either an X- or a Y-chromosome. Whether the offspring is a genetic male or female, then, depends on which kind of sperm cell fertilizes the egg, creating either an XX or XY in the twenty-third pair.

The information needed to make proteins and enzymes in the cells of the body is coded by a molecule of DNA within each chromosome. A remarkable amount of DNA is packed into each chromosome. The DNA extracted from all forty-six chromosomes and laid end to end would extend six feet.[9] The building blocks of DNA are called nucleotides, and there are four types: adenine, thymine, cytosine, and guanine. What geneticists have learned from studying the DNA sequences is that all humans, everywhere on the planet, have exactly the same set of genes. But some of these genes come in slightly different versions. If you compare two individuals, taken at random, you will typically find about two differences in their nucleotides for every one thousand examined.[10] It is from these differences in the DNA sequences of our genes that the differences emerge between the two individuals. An observed characteristic or trait of an individual is referred to as a *phenotype*, whereas a pattern of nucleotides in the genes comprise a *genotype*. Differences in the genotype, then, provide a path for differences in the phenotype. Our skin color, eye color, hair color, head shape, and body dimensions in general are examples of such phenotypic differences. Susceptibility to diseases is another example. One person might be more susceptible to a particular disease than is another person. A disease may be virtually preordained by the genotype, such as Huntington's disease, or the genotype may only increase the chances of contracting a disease, such as a heart attack or stroke. Lastly, some of our behavioral or cognitive characteristics, such as temperament and intelligence, can also be influenced by the genotype. For these in particular, more than one gene is responsible, and the influence of the gene is always probabilistic rather than certain.

Comparisons with Chimpanzees

Phylogeny refers to the evolutionary history of the different kinds of animals or plants. The human genome has been compared with the genome of our closest living primate relatives as part of the study of phylogeny. Strikingly, the genome of the common chimpanzee (*Pan troglodytes*) has been found to be remarkably close to that of *Homo sapiens*. In terms of single-nucleotide substitutions, the two genomes differ by only 1.23 percent; insertions and deletions (3 percent) and duplications (2.7 percent) reveal somewhat greater differences.[11] As a simple summary of the similarity of the two species, human beings share roughly 98 percent of our DNA with chimpanzees.

Although the genes of humans and chimpanzees largely overlap, there are numerous obvious differences in body structure in the two species. Compared with the chimpanzee, human beings are habitually bipedal, larger brained, taller, and less muscled, for instance. The cognitive differences between human beings and chimpanzees are even more profound, as will be argued throughout the remaining chapters of the book. Notably distinctive are the human capacities for planning, thinking strategically, and resisting the impulses of the moment; for imitating and reading the minds of others with respect to their intentions and beliefs; for producing and comprehending language; for interiorizing language and linking it to causal interpretations of perceived events and the actions of others; and for mentally time traveling back into the past or forward into the future to experience events outside the world of immediate perception.

Despite the major phenotypic differences between the two species, evolutionary biologists interpret the strong genotypic similarity as evidence that *Homo sapiens* and *Pan troglodytes* once shared a common ancestor in the tree of life. Approximately five to seven million years ago, it is theorized, the branch of life including the African great apes split into two different smaller branches.[12] One of these eventually led to the development of hominids, including modern human beings. The other branch led to a different set of species that eventually culminated in modern chimpanzees. According to evolutionary theory, it is a fallacy to say that human beings evolved from chimpanzees. On the contrary, each species alive today resulted from a different line of descent. The two species are related only because they presumably shared a common ancestor alive approximately six million years ago and thus inherited many common genes.

The postulated branching of the primate family tree is calculated to have occurred approximately six million years ago by using the small 2 percent difference in genomes to estimate the time elapsed since the divergence from a common ancestor. The calculation starts with a determination of the rate at which mutations occur in the genome, and this rate is assumed to be constant over time. Then, by knowing the current number of DNA differences between the two species, biologists calculate an estimate of the time that has passed since the split from a common ancestor.

The first lesson taken here from the recent discoveries regarding the human genome is that human beings and chimpanzees have followed separate paths for millions of years. It is thus not surprising that the two species differ in body and mind, given the accumulation of six million years of evolutionary changes in each line of descent to the present. One branch of the tree led to *Homo sapiens* and an independent branch led to *Pan troglodytes*, if the inferences from DNA analyses and models of evolutionary biology are correct. The separation and individuality, rather than commonality, of the two species has been underscored by the discovery of fossils from the earliest known hominid species, *Ardipithecus ramidus*.

Dated at 4.4 million years old—relatively close to the hypothesized branching point—the skeletal remains of *Ardipithecus ramidus* provide the best approximation scientists have to the bodily structure of the last common ancestor.[13] *Ardipithecus* may have been an ancestor that led to the genus *Homo* or instead may have been part of a side branch that eventually went extinct. In either case, the fossils offer evidence of the skeletal structure of a hominid that is more ancient than any previously discovered species, and they revealed surprising differences from today's apes. Unlike modern chimpanzees, bonobos, and gorillas, *Ardipithecus ramidus* appeared to be bipedal with a pelvis and foot capable of upright walking. Also, the large projecting canine teeth of male chimpanzees were absent in the earliest known hominid species; instead, the males possessed small canines similar to the females in *Ardipithecus ramidus*. In fact, the overall body size of males was only slightly larger than that of females in early hominids, in contrast to the sexual dimorphism seen in chimpanzees, with males markedly larger than females.

Chimpanzees and their close sister species, bonobos, are the species alive today with the closest genetic relationship to modern human beings. An interest in comparing the mentality of these apes with humans is therefore natural and of great interest in cognitive social neuroscience. Even so, if the calculation of shared ancestry is correct, human beings and chimpanzees are the unique products of six million years of independent evolution. With *Ardipithecus ramidus*, paleontologists get a glimpse of a species removed from the branching point by only two million years or so, and it is apparent that the hominid branch had already diverged considerably from the branch that led to modern chimpanzees. It is thus important to keep in mind the limitations of comparing modern chimpanzees with modern humans. Despite their striking genetic similarity, each is the product of 6 million years of potential evolutionary change.

Out of Africa

A second lesson from the human genome project is that all modern human beings belong to a single species with a single origin. Scientists have proposed two hypotheses regarding human origins and the spread of *Homo sapiens* across the face of the earth. One is the out-of-Africa scenario. Paleontologists discovered a cluster of skulls of early *Homo sapiens* from an earlier time period than those found outside of Africa.[14] The oldest known fossils of our species came from Ethiopia and have been dated to 130,000 years ago. Other early fossils of modern humans were found in a cave near Nazareth in Israel; these are estimated to be ninety thousand to one hundred thousand years old. All fossils of *Homo sapiens* found outside Africa and the Middle East so far have been identified as more recent in origin—no more than about forty thousand years old. One interpretation of these age differences is that modern humans originated in Africa and later migrated to the Middle East and eventually to Asia and Europe.

Genetic diversity provides a clock by which biologists can calibrate the antiquity of human populations. The more time to accumulate mutations, the more differences will be observed among individuals of the population. According to the out-of-Africa hypothesis, the primordial group of *Homo sapiens* grew in population and eventually began to migrate to other locations. The paths of migration can be modeled by comparing the genetic diversity of groups living in various regions of the world today. Put differently, the variations that are found in the DNA of populations around the world today can be traced back to the stock of possible genetic diversity established first in ancient Africa. For example, as human beings migrated to the Middle East from Africa, they carried in their cells the nuclear DNA of their ancestors with the subset of mutations that had accumulated at that point in time. The more diversity found in a population, the longer it has been there. Conversely, the less diversity found, the more recently the population had migrated to the region. By developing models of the patterns of migrations from the genetic evidence, scientists conclude that groups of people migrated from Africa first to other parts of the Middle East. Later, they migrated to Asia and northern and eastern Europe. The Americas were populated much more recently by a migration across the Bering Strait from eastern Asia to the region of present-day Alaska.[15]

The alternative hypothesis argues that there may once have been a migration from Africa, but it was not *Homo sapiens* who migrated but rather an earlier ancestor, possibly *Homo erectus*. From ancestral populations of an earlier ancestor, such as *Homo erectus*, modern human beings evolved independently of one another in multiple regions across the planet. That is to say, modern human beings in Asia evolved from the Asian variant of *Homo erectus* whereas modern Europeans evolved from the European variant. This alternative view is known then as the multiregional hypothesis. The two hypotheses have long competed to explain the skeletal fossils and cultural artifacts discovered in different regions of the world. However, human-genome research offered fresh evidence favoring the out-of-Africa over the multiregional view.

One approach involves analyzing mitochondrial DNA. This is of interest because mitochondrial DNA is passed on from one generation to the next solely from the egg cells of mothers. The father's sperm cells do have a few mitochondria, but to the great fortune of geneticists, all of the sperm mitochondria are discarded in the fertilization process. This means that geneticists can trace the variability in mitochondrial DNA of today's women back to their mothers, which in turn can be traced back to their grandmothers, then their great-grandmothers, then their great-great grandmothers, and so on. In theory, one could reconstruct the family tree backward in time to identify a female who was the source of all mitochondrial DNA existing today, sometimes referred to as mitochondrial Eve. Mitochondrial DNA offered a picture of the family tree undistorted by the blending of maternal and paternal genes that occurs in the DNA of the cell nucleus. Furthermore, mitochondrial DNA mutates at a much faster rate than nuclear DNA, so it can provide a clock capable of tracking comparatively short passages of evolutionary time. The family tree constructed from these data had two major branches—one contained all mitochondrial DNA variations today found only in Africa; the second branch contained a mixture of all types. This outcome was thus consistent with the out-of-Africa theory of human origins.[16]

The variability in mitochondrial DNA is largest among African populations, as would be expected if they have been around the longest and accumulated the most mutations over time. The study of nuclear DNA confirmed this conclusion by showing that of the total genetic variations on a particular chromosome found in populations around the world, most all of them are represented in the African populations. Relatively few were found in Europeans, Asians, and Americans. In sum, Peter Raven and George Johnson concluded, "Taken together with fossils of early H. sapiens from Africa and Israel, these results strongly support an out-of-Africa model of human origins."[17]

The view that the human mind originated with an ensemble of five distinctive parts would have no credence at all if the multiregional view of human origins were correct. It is inconceivable that such an ensemble emerged not just once but independently multiple times in different populations in different geographical locations. Thus, the ensemble hypothesis is worth entertaining only because modern human beings arose and only once.

THE HUMAN BRAIN

The genome shows our common origin as human beings and so does the anatomy and physiology of our brain. Our conceptions of the mind, and ultimately society, can be fruitfully viewed through the lens of neuroscience. That is not to say that scientists have solved the problem of how the mind and brain are linked. Our perceptions, thoughts, feelings, memories, and dreams—the makings of mind—are qualitatively different from the neural and other bodily cells that make up the organ of the brain. Understanding this duality of mind and brain remains a profound philosophical and scientific challenge that we are not yet near resolving. Still, at the beginnings of the twenty-first century, scientists take for granted that the neurobiology of the brain informs our descriptions of the mind and that, in some fashion, the duality must be bridged.

Cognition is assumed to be a function of the brain, just as breathing is a function of the lungs or blood circulation is a function of the heart. The human brain may well be the most complex structure in the known universe. Consider just a few of the brain's properties to understand this point. A neuron includes dendrites for receiving signals from other neurons, a cell body, and an axon for transmitting a signal to other neurons via a synaptic connection. The dendrites of a single neuron may receive thousands or even tens of thousands of synaptic connections from other neurons. With about a trillion (10^{12}) neurons in the brain, there are at least 1,000 trillion (10^{15}) synaptic connections among these neurons.[18]

Structurally, the neurons of the brain are organized in terms of the hindbrain, the midbrain, and the forebrain, each developing from distinct portions of the neural tube that forms after conception. The forebrain eventually divides into two parts during maturation. The first part is known as the telencephalon, with its massive left and right cerebral hemispheres and associated structures that make up much of the human brain. The cerebral hemispheres are visible to the eye on inspection. The associated structures include the basal ganglia that underlie the cerebral cortices and the structures of what is known as the limbic system. Lying beneath the telencephalon is another "hidden" interior region of the forebrain known as the diencephalon, which includes the thalamus and its smaller partner the hypothalamus. All of this rests on top of the brain stem. The brain stem itself is composed of a midbrain (mesencephalon) at the top and, below that, a set of structures known as the hindbrain (the metencephalon and myelencephalon).[19]

The human brain is notable for its large size, especially when compared with the size of the body. The typical endocranial volume of a human being is 1,355 cubic centimeters, and this is about three times the estimated brain size of the hominid species known as *Australopithecus africanus* (457 cm^3), which paleontologists date as having lived 2.6 to 3.0 million years ago.[20] Although the skull size of a modern chimpanzee looks from the exterior to be about comparable in size to that of a modern human, its cranial capacity is much smaller. As Richard Klein explained in *The Human Career*, the typical chimpanzee brain size is about 400 cm^3, or about the same size as that of the ancient genus *Australopithecus*. Within the advent of the genus *Homo*, the fossil record supports the conclusion of an increase in brain size from very early to more recent species, but the "long-term increase in endocranial volume in *Homo* was not due simply to increased body size," as "it was accompanied by a less dramatic, but still conspicuous decrease in the relative size of the cheek teeth (premolars and molars)."[21]

The remarkable increase in cranial capacity documented in the field of human paleontology is arresting. The large brain of modern *Homo sapiens*— and that includes all of its diverse populations—is fundamental to understanding who we are as a species. The massive expansion of the brain in the modern human being undoubtedly was a necessary condition for the ensemble of capabilities posited as unique to the mind within us. If not one but five significant cognitive systems distinguish the human brain, then only a massive increase in size could accommodate all five parts of the ensemble.

The size of the brain relative to body size—what is known as the encephalization quotient—provides a way to take into account that large animals are likely to have large brains and small animals, small brains. Whales, dolphins, and elephants, for example, all have larger brains than do human beings because they have larger heads and bodies in general. A plot of the mean brain and body weights of a large number of primate species on a log-log scale enables one to look at this relationship over a diverse range of body sizes. The plot turns out to be a straight line that shows a very strong positive correlation (+0.97), almost a perfect relationship of 1.00. That is to say, the plot shows this without including *Homo sapiens*. Tellingly, the data point for human beings lies well above those of the chimpanzee and the gorilla. It is in fact about three times higher than it should be according to the straight line of regression found among all other primates.[22]

In addition to the relative size of the human brain, changes also apparently occurred in its organization across the course of hominid evolution. That is to say, the relative size of specific brain regions have increased or decreased as opposed to an overall expansion in size. Comparative neuroscience has revealed several differences among living species.[23] For example, primates dedicate more cerebral cortex to visual functions in comparison to rodents, which show relatively larger areas dedicated to sensing odors rather than seeing. Primates in general have relatively large prefrontal regions that are anterior to the motor areas of the frontal lobe. Among the functions of these prefrontal areas are social behaviors that are known to be complex in great apes and in human beings. The left hemisphere of most human beings is dominant in controlling motor behavior such that we are more often than not right-handed due to contralateral control, with the left brain controlling the right side of the body. Chimpanzees show some similar asymmetries in the frontal and temporal lobes, but they lack the strong degree of asymmetry in handedness.

Yet trying to account for differences in cognitive capabilities between chimpanzees and human beings in terms of brain reorganization—as opposed to sheer brain expansion—has not been straightforward. For example, as will be discussed in chapter 4, language in the human brain depends on a neocortical region in the frontal lobe of the dominant left hemisphere. It is known as Broca's area, and the region is known to be relatively larger than its counterpart in the right hemisphere. If this left-right asymmetry were absent in chimpanzees, then one could reasonably hypothesize that a reorganization of the left frontal lobe played a role in the origin of language in the human brain. Yet neuroimaging studies have documented a similar left-right asymmetry in chimpanzees, bonobos, and gorillas.[24] If there are differences between the language areas of the human brain and their homologous counterparts in the chimpanzee, then it must be in the microanatomy of local neural circuitry or the connections between cortical regions rather than at a more obvious level. The ensemble hypothesis stresses that the reorganization from nonhuman to human of any one cognitive system—such as working memory—might be relatively modest. It is the sum of all five parts and their nonlinear interactions that result in a qualitatively different kind of mind.

Further insight may ultimately come from comparisons of the human genome with the chimpanzee genome. An exciting lead identified regions of the human genome that have dramatically changed from the chimpanzee genome while being conserved without alteration among other mammals in general. These regions are potentially interesting hot spots. Referred to by geneticists as human accelerated regions, they total forty-nine in number (HAR1–49). The one that has changed the most in relation to the chimpanzee genome, HAR1, is part of an RNA gene that is expressed during human cortical development.[25] The finding is especially intriguing because the RNA gene (HAR1F) is expressed while the brain is developing in the womb from seven to nineteen weeks after gestation. This time window is known to be a period when cortical neurons are migrating into position to determine the structure of the six layers of human neocortex. Perhaps it will someday be possible through further neurodevelopmental research to know the details about how the human cortex differs from the nonhuman cortex. It is anticipated that massive cortical expansion, coupled with relatively modest reorganization in the neural networks underlying five key parts of the modern ensemble, will prove vital to the uniqueness of the human brain.

So far, the emphasis has been on how the human brain might differ from nonhuman species. The chapter will now turn to the remarkable commonalities between human and nonhuman brains. Not only do all primates follow the same basic plan for the brain, but so do all vertebrates. The hindbrain, midbrain, and forebrain organization is a constant. At the level of the individual neuron, a similar constancy across life forms can be seen. The pioneering studies of how nerve impulses are conducted in the mammalian nervous system used the neurons of squid, horseshoe crabs, and frogs. The most basic nervous system—a simple net of neurons—is found in the jellyfish. Yet the same electrochemical mechanisms at work in the jellyfish are also at work in the human brain.[26]

Besides being able to communicate nerve impulses from one cell to the next, all primates, all mammals in general, and even reptiles, must send signals from the brain to the body via the spinal cord and, conversely, receive messages in the brain from the body. The most primitive parts of the brain, then, are in the metencephalon and myelencephalon of the hindbrain, with direct connections to the spinal cord. A key function of the hindbrain is neural control of the lungs to breathe and the heart to circulate blood, essential life support mechanisms.[27] These, as well as many other essential functions, are just as necessary for life in fish and amphibians as they are for life in reptiles and mammals. It is as if nature conserved brain components that worked well—from individual units such as neurons to complex structures such as brainstems—and reused them in other species.

The mesencephalon or midbrain includes a nucleus of cells known as the substantia nigra, which produces a neurotransmitter called dopamine. A loss of the functioning of these neurons results in the motor control difficulties of Parkinson's disease.[28] Dopamine is also a central neurotransmitter in the reward circuit of the brain, which plays a major role in drug addiction, as will be seen in chapter 6.

Above the brainstem in the forebrain, however, significant differences appear in different classes of the animal kingdom. Mammals are characterized by more complex forebrains compared with reptiles. The diencephalon of the forebrain is the interior, lower region that lies on top of the brainstem and includes the thalamus and a smaller structure, the hypothalamus, which connects with the most primitive portion of the telencephalon, known as the limbic cortex or rhinencephalon. This primitive structure makes up virtually the entire forebrain of the crocodile. As Richard Thompson clarified, "No negative inferences about the function of the limbic forebrain or of crocodiles is meant, for in addition to being vicious, the crocodile is an intact functioning organism responsive to sensory stimulation and engaging in a variety of behaviors: feeding, fleeing, fighting, and reproduction."[29] The hypothalamus is linked with several limbic structures important in memory (hippocampus) and emotion (amygdala) that will be discussed in chapters 6 and 7. Together they constitute the limbic system that plays a central role in our feelings of pleasure, happiness, anger, fear, sadness, and other human emotions.

The other part of the diencephalon, the thalamus, relays signals to higher regions of the brain in the most advanced layer of the telencephalon, the cerebral hemispheres.[30] These inputs to the cerebral hemispheres from the thalamus regulate the brain's state of wakefulness and activity level from signals originating in the brain stem. They also convey the inputs from all our peripheral sensory systems, such as vision, taste, and touch, to the sensory regions of the telencephalon; these are located in the outer neural covering of the cerebral hemispheres called the neocortex. The neocortex can be divided into four general regions. The frontal lobe is the most anterior region behind the eyes and forehead and extending back to a dividing central fissure. Just posterior to the central fissure is the parietal lobe. Behind the parietal lobe at the very posterior end of the cerebral hemispheres is the occipital lobe. The lateral fissure divides off the temporal lobe—near the ears—as the region inferior to the frontal and parietal lobes. Much of what will be discussed in later chapters regarding the ensemble hypothesis makes reference to these specific regions of the cerebral cortex.

The neocortical regions are much larger in the forebrain of human beings compared with other mammalian species. Paul MacLean theorized that the human brain is a composite of the expanded neocortical areas and phylogenetically older regions of the limbic system and the diencephalon.[31] Of special importance, he proposed a conception of the limbic system as a brain network shared in common by all mammals and presumably inherited from the very earliest mammals in evolutionary history. The hippocampus, amygdala, and other structures constitute a border surrounding the diencephalon and were referred to as the limbic system. The emotional feelings provided by the limbic system aided the survival and reproduction of individuals. On top of this older limbic structure is the neocortex, which enables the brain to take in visual, auditory, and other perceptual information about the environment to make reasoned, unemotional decisions that benefit survival. In humans, the neocortex is so large and well-developed that it can support reasoning, problem solving, language, and even literacy in the form of reading and writing.[32] In terms of the argument here, the five parts of the ensemble hypothesis resulted primarily from the expansion and reorganization of the neocortical forebrain.

As knowledge in the cognitive social neuroscience progressed, Maclean's view that the limbic system was dedicated to emotional processing proved too limited. It became clear that a key part of the limbic system, the hippocampus, plays an important role in the cognitive function of memory storage in nonhumans as well as in humans. It also became clear that the human experience of emotion depends on activation of regions in the neocortex as well as on limbic structures. Still, Maclean's hypothesis appropriately stressed the composite nature of the human brain. As Joseph LeDoux summarized it in the *Annual Review of Neuroscience*: "In particular, the notion that emotions involve relatively primitive circuits that are conserved throughout mammalian evolution seems right on target," as is the idea that "cognitive processes might involve other circuits and might function relatively independently of emotional circuits, at least in some circumstances."[33] Maclean also accurately described the massive expansion and reorganization of the forebrain as the neural underpinning of the modern human mind.

OVERVIEW

The following chapters address the unique mental capacities—such as executive function, language, and mental time travel—that make for a beautiful mind, not in only a few creative geniuses, but in us all. These capacities enable our most uplifting and laudable characteristics as human beings, our morality and spirituality. Yet they also mediate the most shocking evil. We alone are capable of murdering—with intent and malice—entire populations of other human beings. Our capacity for genocide perplexingly coexists with our moral capacity to know that such acts are wrong. Each such defining feature, the good and the atrocious, sets human beings apart from all other inhabitants of earth.

In chapter 2, the development of an advanced form of working memory is considered. The key ingredient is the improvement of the executive functions of working memory that allow for human beings to plan ahead, to inhibit impulses, and to delay gratification. However, other changes also occurred in working memory as an interaction with the advent of language. A specific store for verbal information was added in addition to temporary storage components for visual and spatial information. These changes together provided for a flexibility in human behavior and the capacity to innovate and plan new ways of doing things. Without these changes, the "human revolution" in creating cultural artifacts would never have started.

In chapter 3, the advanced social intelligence of human beings is considered. Because of its power, human beings moved beyond the glacially slow form of change found in natural selection and modification by biological descent through genetic inheritance. Changes in the human mind over the history of modern human beings have instead occurred through the rapid process of cultural evolution. Unlike any other species, the human brain, from the moment of birth, is immersed in a complex culture that profoundly shapes the functioning of the mind. The human mind cannot be reduced to a genetically predisposed brain structure because the culture in which we are raised is so rich and important to mental development. The human brain includes mechanisms for understanding the mental intentions of others and for learning from the behavior of others through imitation and direct instruction. These mechanisms—interacting with advanced working memory—paved the way for the cooperative invention of new artifacts and the social transmission of culture.

In chapter 4, symbolic thought and its extension in language is addressed. Of the symbols that inhabit our cultural mind, words are the most mysterious. Arbitrary sounds emitted by the human vocal tract stand for objects and events in the physical world and abstract ideas in the world of thought. By stringing words together, speakers of the same language can communicate everything and anything. This is possible only because the left hemisphere of the human brain has become specialized for the production and comprehension of speech. This distinctive part of the ensemble accounts for why there has been only limited success in attempts to teach chimpanzees language. That we can communicate about abstract ideas in essentially an infinite variety of ways is rightly viewed as fundamental to the divide between human and ape mentality.

As described in chapter 5, a fourth part of the ensemble emerged from an interiorized form of language—the inner voice—combined with an ability to make causal inferences. Instead of using language to communicate with others, it migrated inward as a medium for silent thought. The inner voice capitalized on our advanced working memory to allow for a running narrative of the everyday experiences that enter our conscious awareness. Trains of connected thought can run, sustained by our verbal working memory. Moreover, this inner voice teamed up with a capacity to infer the causes of events even when they are hidden and not perceptually obvious. This combination of interiorized language and causal inference provided the human mind with a new cognitive tool to interpret conscious experiences. This interpretive function of the left hemisphere has been discovered in research with patients who had undergone surgery to separate the left and right hemispheres as a means of controlling severe epileptic seizures.

The human mind is the only known exception to the unidirectional flow of time in nature, as will be explained in chapter 6. To recollect the what, when, and where of past events implies our capacity for mental, if not physical, time travel. Not only can the mind go back in time to relive past experience, but it can also use the same brain network for moving forward in time to imagine future events. Much of what makes the human mind so unique stems directly from our recollections and replays of the past and our fantasies and apprehensions about the future. The hazards as well as the benefits of mental time travel are outlined in the chapter.

In sum, chapters 2 through 6 specify each of the five parts of the ensemble that distinguishes the modern human mind. In the *Descent of Man and Selection in Relation to Sex*, Charles Darwin stressed the underlying continuity of the human mind with the nonhuman as follows:

> There can be no doubt that the difference between the mind of the lowest man and that of the highest animal is immense. An anthropomorphous ape, if he could take a dispassionate view of his own case, would admit that, though he could form an artful plan to plunder a garden—though he could use stones for fighting or for breaking open nuts—yet that the thought of fashioning a stone into a tool was quite beyond his scope. Still less, as he would admit, could he follow metaphysical reasoning, or solve a mathematical problem, or reflect on God, or admire a grand natural scene. . . . Nevertheless, the difference, great as it is, certainly is one of degree and not of kind.[34]

According to the ensemble hypothesis, Darwin was wrong here. The immense difference between the modern human mind and any nonhuman mind is in fact a matter of kind and not degree. Darwin was certainly aware that language by itself seemed to represent a discontinuity, for he went on to state: "If it could be proved that certain high mental powers were absolutely peculiar to man, which seems extremely doubtful, it is not improbable that these qualities are . . . mainly the result of the continued use of a perfect language."[35] He, of course, was unaware of the findings of the cognitive social neurosciences that began to accumulate about one hundred years after he first published *The Descent of Man* in 1871. The continuity thesis of Darwin in the realm of the mind—if not the body—failed to appreciate the summative effects of five different parts of the modern mental ensemble. Further, and this is fatal for Darwin's view that we differ only as a matter of degree from apes, it failed to recognize the nonlinear interactions of these parts. The pairwise and higher order interactions of advanced working memory, advanced social intelligence, the left-hemisphere interpreter, and mental time travel, as well as symbolic thought embodied in language, produce a qualitatively different kind of mind.

In chapters 7 and 8, it is shown how the five parts of the ensemble hypothesis are revealed in human emotion and our readings of the minds of other people in social relationships. Both emotion and a social nature are, of course, found in all primates, and in mammals more generally. Yet it is contended here that human emotion and the human social mind are thoroughly colored by the modern mental ensemble. At the pinnacle of human mental functioning perhaps are our capacities for morality and spirituality. These, it is argued in chapters 9 and 10, are again largely, if not entirely, products of the ensemble.

Finally, how is the human mind changing in the twenty-first century? Despite that it is biologically the same as it was forty thousand years ago, cultural evolution has dramatically altered the way the modern mind functions. As argued in chapter 11, the emergence of the literate mind through the invention and multiple uses of writing must be seen as a product of the modern mental ensemble. The ways literacy altered human cognition over the past few thousand years of history were enabled by specific parts of the ensemble. Today we find ourselves both exhilarated by and drowning in the information overload of the telecommunications and computer revolution of the late twentieth century. Just as the ensemble produced a literate mind from the invention of writing, so, too, will it shape a postliterate, Internet mind from today's cognitive technologies. The book closes, then, with a look toward where our capacities to plan, to collaborate, to write, to interpret, and to imagine might someday take us.

Chapter 2

EXECUTIVE WORKING MEMORY

The human mind is highly adept at executive functioning—the ability to make decisions and manage other cognitive processes. Our ability to plan solutions to novel problems—as we have been doing as a species since the dawn of humanity in the Upper Paleolithic—is a common indicator of executive function. Another is the ability to delay immediate gratification to obtain a more valuable reward later (e.g., work hard now in order to get paid at the end of the week). A third example is the ability to inhibit impulses that offer pleasure but conflict with other important personal goals (e.g., skip the donut with morning coffee to pursue a goal of losing weight). Executive function is known to require the successful maturation of the prefrontal cortex of the brain during infancy, childhood, adolescence, and even extending into young adulthood. This is because the prefrontal cortex includes a neural network serving executive attention and other components of working memory. This system of short-term memory enables the brain to retrieve mental representations from long-term storage and maintain them in an activated state. Our ability to think before we act depends on this transient form of memory. Facts and skills that have been learned and stored in long-term memory can be retained in an inactive state for hours, days, weeks, months, or years. By contrast, working memory stores information for only half a minute or so. If executive attention is focused on the words, images, and other thoughts, then these mental representations can be kept active for a few moments longer within our immediate conscious awareness. Without continued attention, however, the contents of working memory dissipate rapidly and slip out of mind.

Try to multiply 8 × 14 without writing anything down. The numbers must be maintained in working memory and attention given to multiplying 8 × 4. By storing a 2 in the digit column and carrying a 3 in the tens column, one can then proceed with multiplying 8 × 1 and adding the carried 3. The correct answer will emerge from this mental work only if working memory can effectively deploy its capacity for attention and transient storage of mental representations. In a similar way, solving novel problems, delaying gratification, and inhibiting impulses require the effective deployment of executive attention and the short-term storage of working memory. Holding in mind alternative possible solution paths while problem solving, comparing the value of immediate to delayed rewards, and focusing on a long-term goal over immediate pleasure illustrate some of the reasons why this is so.

Because working memory is transient, its value in human cognition may be less apparent than the value of long-term memory, the repository of everything learned throughout a lifetime. It is obvious that human and nonhuman intelligence requires an ability to learn and retain things for many years. Yet bringing these things into awareness and holding them as long as they are needed is also essential. A leading researcher of the neural basis of working memory, P. S. Goldman-Rakic, explained its real value:

> The brain's working memory function, i.e., the ability to bring to mind events in the absence of direct stimulation, may be its inherently most flexible mechanism and its evolutionarily most significant achievement. At the most elementary level, our basic conceptual ability to appreciate that an object exists when out of view depends on the capacity to keep events in mind beyond the direct experience of those events. For some organisms, including most humans under certain conditions, "out of sight" is equivalent to "out of mind." However, working memory is generally available to provide the temporal and spatial continuity between our past experience and present actions.[1]

The flexibility in thought and behavior afforded by working memory led some scholars to wonder about its role in the great leap forward in tool making and symbolic art associated with the origin of the modern human mind. Possibly a genetic mutation produced a reorganization of the brain compared with other members of the genus *Homo* that resulted in an advanced form of working memory. Consider that stone tools made by *Homo neanderthalensis* in the cave shelters of Europe included scrapers, sharp points, and tools with thin edges. The design of these tools represents the Mousterian style; it took considerable skill to craft the stone for the functions of scraping and sawing or to attach the sharp points to a spear for use in hunting. The Neanderthals were obviously capable of acquiring the expertise needed to make such tools and perhaps did so in the same way work skills are acquired still—through practice and apprenticeship with an expert craftsman. It seems likely that the knowledge and skills were passed on from one generation to the next by the young novice learning from the old master. However, it is curious that the stone tools of the Neanderthals did not appear to change and improve in design over the two-hundred-thousand-year-long history of the species.[2]

Despite the impressive tool-making skills of the Neanderthals, archeologists have not detected signs of innovations over multiple generations, as is so characteristic of the cultural evolution of modern humans. Even from one generation to the next, our species may make improvements in the design of a tool or product and pass it on to the next generation to mull over and refine, such as innovations in the forms of wheeled transportation from the chariot to the contemporary car. Yet, for more than two hundred thousand years, Neanderthals appeared to make the same model of scrapers, for example. They continued to make, without tinkering or planning new designs, the same kind of stone-tipped spears for hunting. In the same vein, carved figurines and representational cave paintings of animals—both highly inventive and planning intensive—are associated only with modern humans.

To illustrate, consider a famous piece of representational art made by modern humans, the Hohlenstein-Stadel figurine.[3] It was discovered in southwestern Germany and dated as thirty to thirty-three thousand years old. The figurine, carved from ivory, represents the body of a man and the head of a lion. Such chimerical beasts have been part of the folklore, mythologies, and religious practices of modern humans from our origins in prehistoric times. The key point here is that to carve a chimera its creator needed to hold in mind two different concepts at once and then combine some of the properties of a lion with some of the properties of a man. The visual and spatial representations of each concept had to be kept active in working memory, while executive attention was directed to the head of one and the body of the other. The fascinating art work of the Hohlenstein-Stadel figurine provides insight into the advanced working memory capacity of early modern human beings.[4] Gaining the power of an advanced system of working memory, then, could be the cognitive Rubicon crossed only by modern human beings in the course of hominid evolution.

Another clear example of the executive functioning of modern human beings is the use of lines of mounded stones as corrals for herding and then trapping gazelles. Evidence of such ancient hunting technology dates back only twelve thousand years or so, indicating they were the design of modern humans and required "the delayed gratification and remote planning in space and time that is typical of modern executive functions."[5]

Cognitive psychologists have developed and empirically tested detailed theories about the cognitive architecture of the modern human mind. These theories describe the basic functional systems of human cognition, such as the distinction between working memory and long-term memory. The objective of cognitive neuroscience is to map these functions onto the structures of the brain outlined in chapter 1. More precisely, the goal is to map particular functions, such as working memory, on specific neural networks that may be localized in multiple brain regions and structures. The nervous system, as well as the respiratory system and the cardiovascular system, are structured in a hierarchy of parts and subparts. For example, the nervous system consists of a peripheral branch and a central branch, which in turn breaks down into the spinal cord and the brain. The peripheral branch includes sensory and autonomic components. The latter in turn breaks down into the sympathetic and parasympathetic parts. In a similar way, the cognitive architecture must be understood in terms of a hierarchical structure.

An influential model of working memory was proposed by Alan Baddeley in his 1986 book titled *Working Memory*. It included two storage components—the phonological loop and the visuo-spatial sketchpad—and a central executive that supervised the contents of these stores through attention. That is to say, executive attention retrieved information from long-term memory and manipulated it in one of the stores, such as occurs during mental arithmetic. The storage components are tailored to specific kinds of information. One type extensively investigated by cognitive psychologists is dedicated to the storage of verbal information in the form of the sound or phonology of words. Baddeley referred to it as the phonological loop to emphasize how verbal information can be recycled by silent mental repetition or rehearsal, as we might do with the name of a person to whom we were just introduced. By keeping the name active in verbal working memory, we can address the individual by name moments later. Besides words, visual-spatial representations can also be stored in working memory. It is this storage device that is called into service in the experience of visual imagery. Imagine the living room of your current residence. Such mental imagery can be maintained for several moments within the visuo-spatial sketchpad. This sketchpad actually consists of two independent stores, one for visual representations and another for spatial locations. Whereas the visual store makes use of neocortical regions in the left hemisphere, the spatial store depends on a neural network in the right hemisphere.[6]

Long-term memory is also hierarchically organized. A major division separates declarative knowledge of facts, concepts, and specific events learned in the past from nondeclarative or procedural knowledge of skills and conditioned habits. It divides our "what" knowledge from our "how" knowledge. For example, long-term memory stores "what" a bicycle is separately from "how" we ride a bicycle. The "what" storage entails a concept of bicycles in general or perhaps a memory of an event with a specific bicycle (e.g., our childhood bicycle seen for the first time one Christmas morning). The "how" storage is very different. Instead of a verbal description or visual image, the procedures for riding a bicycle are remembered in the muscles and senses as a perceptual-motor skill. An amnesic patient might forget that he ever owned a bicycle, but he would still retain the skill of riding one because the two kinds of memory are stored independently of each other.

Another way to think about the organization of long-term memory cuts across the distinction between declarative and procedural storage. The domain of knowledge defines the organization rather than the difference between "what" and "how." Bicycles and all other inanimate objects made by human beings constitute a domain of knowledge that is stored separately from our knowledge about plants and animals as living things. Knowledge about people and social relationships is a third domain. Each of these domains includes declarative knowledge of facts, concepts, and specific events from the past as well as procedural knowledge. For instance, within the domain of knowledge about people, the brain can store the general concept of a human face, a multitude of facial images of specific people met in the past, and the perceptual skill of rapidly recognizing a face from all other objects or a specific face in a room crowded with many people.

Thus long-term memory is organized in terms of distinct domains of knowledge, and each domain is further broken down into specific modules. A module is a system, involving one or more regions of the cerebral cortex, that specializes in a particular domain of cognitive activity. The domain of knowledge about people includes multiple modules dedicated to specific tasks. The most extensively investigated is a perceptual module dedicated to the recognition of faces. Due to damage to the face recognition module, prosopagnosic patients are unable to recognize the faces of people—even the faces of those whom they know very well and can recognize from, say, their speech. Prosopagnosia involves regions lying at the juncture of the temporal and occipital lobes of the brain. The superior temporal sulcus and an area in the fusiform gyrus of the temporal lobe join forces with the inferior occipital cortex to process faces.[7] These neural regions in the right hemisphere are particularly sensitive to faces. Objects must possess the right features, such as hair, forehead, eyes, nose, mouth, and chin, and these features must be configured properly (e.g., nose above the mouth) to elicit a strong response from the right temporal-occipital network. Non-faces, such as inanimate objects like bicycles, are processed by separate neural networks in different, more anterior regions of the temporal lobe.

Finally, the output of the face module is shallow and limited to only seeing the perceptual structure of a face. The Italian artist Arcimboldo painted a series of composite faces in which common objects made up the facial parts. In viewing such paintings, we become aware of both the configuration of the face and the nonfacial component objects. This kind of double awareness occurs because the output of the face recognition module and the output of the system responsible for object recognition both gain access to consciousness. In patients suffering from selective damage to the module that performs the recognition of common objects other than faces, a different outcome takes place. When shown an Arcimboldo painting, such a patient is fully aware that it looks like a face, because the intact face recognition module automatically provides its output to consciousness. However, the face module cannot see any deeper to see the identity of the objects used as facial parts. Without an intact object recognition module working in tandem with the face recognition module, the Arcimboldo painting loses its intriguing effect.[8]

Modules are thought to be biological adaptations that were selected for during human evolution and are now hardwired in the human genome. It is easy to see why face recognition would be important enough to warrant a dedicated perceptual network. Similarly, there is strong evidence for a speech recognition module that rapidly and accurately identifies the short bits of sound (i.e., the phonological segments or phonemes) that are the building blocks of words in spoken language. These perceptual modules have been theorized to feed into more abstract cognitive modules that further organize knowledge within long-term memory.[9] Knowledge about people includes cognitive modules for processing facial expression, language, nonverbal behaviors, and the mental states of others. It also includes modules for organizing knowledge about the self. Another important set of cognitive modules in the domain of social information concerns groups of people. For example, recognition of kin and appropriate ways of behaving toward those biologically related to us is one such cognitive module. Other cognitive modules are dedicated to knowledge about the in-group, the out-group, and other dimensions of group identity such as ideology. Cognitive scientists are still debating how many modules exist within the system of long-term memory. The cognitive modules that represent knowledge about people are certainly only a subset. Biological knowledge about plants and animals and physical knowledge about climate, landscapes, and other ecological features are also important for obtaining the resources needed to survive and reproduce.

Whereas modules are narrowly tailored to the needs of a specific domain, working memory is the general-purpose brain system. It can combine knowledge stored in separate domains of long-term memory on a temporary basis. Consider again the Paleolithic artwork of the Hohlenstein-Stadel figurine. The module representing knowledge about the shape of the human body is stored separately in long-term memory from knowledge about animals, such as lions. The knowledge modules of Neanderthals were possibly as detailed as those of modern humans when it came to such fundamentally important domains as people and lions. But Neanderthals apparently never thought to combine the body of a man with the head of a lion in creating a work of art. This combination required a degree of executive attention and possibly visual and spatial storage capacity in working memory unavailable to them. The advanced form of working memory associated with modern human beings provided a general workspace where the modules of diverse domains could be brought together in innovative ways.[10] The artist was capable of combining knowledge from social modules representing knowledge about people with knowledge about animals as represented in biological modules. The man-lion can be seen as a symbolic representation of an abstract idea, such as the idea of a lion that has taken on human features or vice versa. Either way, the Hohlenstein-Stadel figurine illustrates that modern humans were able to combine in a common workspace multiple domains of knowledge to create artifacts with symbolic meaning.

Similarly, Neanderthals were able to master the procedural knowledge of how to make a stone scraper versus a spear point. A cognitive module for tool making was apparently well established in the Neanderthal mind as part of their physical knowledge of the environment. They could acquire expertise through learning and apprenticeship and store these procedural skills in long-term memory. However, Neanderthals seemed to lack what modern humans possessed: the advanced working memory—particularly executive attention—needed for inventing and planning new ways of doing things. Without innovation, there was no possibility of improving the design of tools from one generation to the next in Neanderthal culture, and this could account for the static archeological record regarding tool design.[11]

EVIDENCE FOR CONTINUITY

As will be described in the last section of the chapter, there are abundant reasons for the claim that the prefrontal cortex of the human brain endows us with a remarkable network of executive attention. Through it, modern humans can, with some effort, inhibit impulses that run counter to our goals for staying healthy or succeeding at school or work. We can delay the need for immediate gratification in order to obtain a superior reward at a later time. We can plan solutions to novel problems. Yet, according to the ensemble hypothesis, these enhancements of working memory were only part of the total story of the human revolution during the Upper Paleolithic. So before turning to characterizations of the power of human executive attention, the evidence for continuity in working memory with other primates will be outlined.

The prefrontal cortex refers to the portions of the frontal lobe that lie in front of the premotor and motor cortex. These motor regions, as well as perceptual regions of the occipital lobe, temporal lobe, and parietal lobe, all provide inputs to the prefrontal cortex. So, too, do the limbic regions and other subcortical structures. Because location and connections matter in the brain, just as they do in the human social world, the prefrontal cortex is positioned "to coordinate processing across wide regions of the central nervous system."[12] The size of the frontal lobe and in particular the prefrontal cortex is quite small in other species compared with human beings. In cats and dogs, for example, the prefrontal cortex is but a small fraction of the total frontal lobe. This disparity in the proportion of the frontal lobe dedicated to prefrontal cortex is also seen in comparing the brain of a rhesus monkey with a human brain. Strikingly, the prefrontal cortex makes up about half of the human frontal lobe; this lobe in turn constitutes about a third of the cerebral cortex in humans.[13]

It is well-established that the human frontal cortex is large and complex. Yet it is not obviously discontinuous with the frontal cortex of the species most closely related to human beings at a genetic level of comparison. Although a reorganization of the brain is apparent when comparing human beings to non-primate mammals, such as dogs and cats, and to primates, as represented by monkeys, the great apes present a different picture. When magnetic resonance images are taken that allow a precise measurement of the size of the frontal lobe relative to the overall size of the brain, continuity is apparent in comparisons of the brain of a human (36.4–39.3 percent) and the brain of a chimpanzee (35–36.9 percent), orangutan (36.6–38.7 percent), and gorilla (35–36.7 percent). All of these species differed from the lower percentage of total cortex allocated to the frontal lobe in the gibbon, a lesser ape (27.5–31.4 percent), and in monkeys (29.4–32.3 percent).[14] These findings contradict older studies that may have suffered from looking at too few or too limited a range of brain sizes in chimpanzees and humans. Moreover, the research tentatively suggested that the prefrontal region "is as large as expected for an ape brain of human size, and that individual humans and individual great apes (but not lesser apes or monkeys) overlapped."[15] Thus, it is harder than was once thought to pinpoint a reorganization of the frontal lobe in general and even the prefrontal cortex in particular that resulted in the superior executive function in human beings.

Continuity between human beings and the great apes can also be seen in a comparison of the storage capacity of working memory. It has been firmly established that the capacity of working memory in modern humans is strictly limited to four items or chunks of information. It had once been thought that human beings could store about seven chunks of information—the length of a telephone number—in the phonological loop. Yet careful investigations revealed that this higher estimate reflected the benefit of rehearsing some of the items and of retrieval from long-term memory. Once a pure measure of working memory capacity is taken, without contamination from long-term memory, a surprisingly small and strict limit of only four chunks is observed.[16] For example, suppose you are trying to keep track of the names of new acquaintances just introduced to you at a party. At the very most, you will be able to retain four such first names within the capacity of the verbal store of working memory. The only way to do better than that is to rehearse some of the names silently in an effort to learn and store them in long-term memory or to use some other mnemonic technique. Obviously, such a verbal test cannot be done with great apes because they lack language and a specialized store for maintaining names. However, they share with us the visuo-spatial sketchpad for maintaining images of objects and scenes. Because the limit of four chunks also applies to the visuo-spatial sketchpad, it is possible to devise a test that directly compares human and nonhuman species.

The Corsi Tapping Test is a common way of assessing the ability to retain nonverbal information for a short period of time. A set of identical icons appears on a touch screen of a computer. The icons then flash or change color one at a time in a sequence. The test taker must then reproduce the sequence by touching the icons on the screen in the correct serial order. One study adapted this test for baboons by rewarding the animal with food when it correctly recalled the sequence by touching the items on the screen in the right order.[17] The number of items in the sequence was varied. Two baboons were tested and both were about 70 percent accurate for three items and only about 20 to 30 percent accurate for four items. A single item more, using a five-item display, resulted in near zero accuracy. For human beings, by comparison, the accuracy rate was 97 percent, 92 percent, and 78 percent. Whereas human beings were clearly better at the task overall, they made errors relatively often when more than four items needed to be retained. For the baboons, this limit was three items instead of four. So the capacity of the visuo-spatial sketchpad in baboons is a bit more limited than in human beings, but the difference is by no means massive.

Finally, the studies that initially pointed to the prefrontal cortex as the site of working memory were performed with monkeys using a different nonverbal test.[18] In a spatial delayed response task, the animal must remember which of two locations contains a food reward. The experimenter baits one of the two locations by slipping food into a covered tray and then blocks the view of the two trays during a short delay period. The monkey gets the food if it can remember where the food is located after it has been out of sight for a while. Electrodes are surgically implanted into the brain of the animal to allow the recording of neural firings while the working memory task is being performed. Such work revealed that neural activity in the dorsolateral prefrontal cortex is central to accurate performance in the task. The important point is that the analogous brain region in human beings has been implicated in spatial working memory. In other words, primates of all kinds rely on the same prefrontal cortical regions to retain the spatial locations of objects over short periods of time.

THE PHONOLOGICAL LOOP

Although the human frontal lobe is not disproportionately large, given the overall size of the brain, it is indeed massive. On average, its volume measures five to six times the volume of the frontal cortex of a chimpanzee; the same overwhelming size advantage applies in comparing the human prefrontal cortex to that of chimpanzees.[19] What use do we humans make of all this cortical space? If we use the metaphor of a house with, say, four times the square footage of another smaller house, then how has the added space been allocated? One option would be to add a room with all the extra space. Perhaps what had been a small entry space just inside the front door could be expanded into a separate parlor, for instance. In other words, the new extra space could be used to add a new room altogether rather than just making the entry way and the living room proportionally larger. If a new room is thought of as a new cognitive function, then it is instructive to use this metaphor, for the human brain is unique in dedicating cortical areas for a verbal store of working memory specialized for learning and using language.

As adults we rely on the phonological loop for comprehending complete spoken sentences. It is also used as you silently read these words, providing an inner voice that accompanies the eye movements of reading. The phonological loop is also important in the production of speech and writing. However, at the most fundamental level the phonological loop is arguably an adaptation that allows very young children to learn words, particularly during the first few years of their lives.[20] By allowing the phonological representation of a novel word to linger in the mind's ear, so to speak, the child can more readily learn to produce the new word and to link the word with its meaning.

Vocabulary acquisition is an enormously important task for the developing child—it is in fact essential for acquiring the culture of one's parents, family, and community. Learning to communicate through language of some sort—spoken or signed—may be the single most important factor in a child's survival and success in life. For children with normal hearing who are immersed in a culture of the spoken word, the acquisition of their native tongue comes easily and at a remarkably fast pace in terms of vocabulary as well as grammar. For word learning, the "rate of acquisition increases during infancy, so that by the age of 5 years, mean vocabulary is in excess of 2000 words. Peak rates of vocabulary growth occur during the school years . . . typically . . . on average 3000 words every year."[21] When children are assessed for their capacity to store information in the phonological loop, they are asked to remember a series of random digits or to try to repeat back nonwords. The digit span test is like trying to remember a new phone number, for example. The repetition test uses nonwords to prevent the reliance on meaning or any other representation of an actual word stored in long-term memory. Instead one must remember phonological segments or syllables that are similar to words but do not fit the pattern of an actual word (e.g., *woogalamic*). From subjects ages three to eight years of age, researchers have found statistically significant correlations between a child's vocabulary size and the ability to perform the digit span and nonword repetition test.[22] Also, when adults try to learn the words of a foreign language in a laboratory setting, their vocabulary learning is impeded when the unfamiliar words sound very similar and are easily confused within the phonological loop. Other experiments injected irrelevant sounds into the phonological loop by having the learner silently repeat the word "the, the, the" over and over again—this suppresses silent articulation and is a good way to distract the loop from accurately storing the sounds of a new foreign word. Learning foreign vocabulary is much impaired when adults are prevented from silently rehearsing each new word in the phonological loop.[23] The evidence from both children learning their first language and adults learning a foreign language shows that the phonological loop has an important cognitive function as an aid to language learning.

The phonological loop is by far the most intensively researched and best understood component of working memory. Neuroimaging studies have been able to identify the regions of the left frontal lobe that support the loop's function in storing and rehearsing verbal material. Broca's area (in the prefrontal cortex) and areas in the premotor and supplementary motor cortex are involved, as is another posterior region in the left parietal lobe.[24] The frontal areas support the rehearsal of phonological representations that are actually stored in and retrieved from a region in the parietal lobe. This network, then, activates multiple regions in the left hemisphere of the brain to facilitate the learning of language. Although the learning of new words may be the reason the loop proved adaptive, its value to language use did not stop there. Cognitive research has also demonstrated its role in silent reading and in the production of speech and written language. As will be seen in chapter 5, the phonological loop is not just for communicating via language. Rather, the inner voice made possible by the phonological loop led to a profoundly important addition to the human mind when it became a medium for silent verbal thought.

HUMAN EXECUTIVE ATTENTION

Besides the unique power of a phonological loop, the mind within us is capable of solving novel problems, delaying gratification, and inhibiting unwanted impulses. Although we may not always succeed in these endeavors, our brain provides us with the potential for success through its massive prefrontal cortex. Using again the metaphor of a house, extra square footage can mean larger rooms or additional rooms, as seen with the addition of a phonological loop. Both of these images seem applicable to the executive functions endowed by the prefrontal lobes of human beings. However, another metaphorical twist is also instructive—the extra space within each room would allow for more furnishings. The executive suite of the human brain is indeed impeccably furnished in the form of superior neural interconnectedness. A striking fact of reorganization at the level of neural circuitry emergent in *Homo sapiens* is the myelination of the subcortical white matter of the brain. Myelin insulates the axons of neurons, allowing for much faster transmission of neural impulses. Comparative neuroscience has clearly documented that the "relative volume of white matter underlying prefrontal association cortices is larger in humans than in great apes . . . compatible with the idea that neural connectivity has increased in the human brain."[25]

The different lobes of the human neocortex do not mature at the same rate.[26] Myelination occurs earliest in the primary motor and sensory cortices but is most prolonged in the prefrontal cortex. Following the same principle, the formation of new synapses is delayed in the prefrontal cortex; the peak synaptic density occurs at about fifteen months of age instead of much earlier for the sensory auditory cortex in the temporal lobe. Unneeded synapses are then eliminated in these neocortical regions slowly over a period of years. Yet again, it is the prefrontal cortex that takes its time with this pruning. Whereas the auditory cortex has ended synapse elimination by age twelve, the prefrontal cortex is still removing synapses to improve the effectiveness and efficiency of its operations into the years of mid-adolescence. In fact, the executive functions of the prefrontal cortex are still developing during the years of young adulthood.

To illustrate the late development of the prefrontal cortex, consider the results of a neuroimaging study using magnetic resonance imaging (MRI) that contrasted the brains of adolescents who were twelve to sixteen years of age with those of young adults aged twenty-three to thirty years.[27] The images showed very little difference in the occipital, parietal, and temporal lobes. It was only in the frontal lobes that postadolescent maturation was obvious. The young adults had maturational changes in the frontal lobes that were from two to five times as great as those detected in other parts of the brain. There was a reduction in gray matter in the young adults compared with the adolescents that "probably reflect[ed] increased myelination in peripheral regions of the cortex that may improve cognitive processing in adulthood . . . for such functions as response inhibition, emotional regulation, planning and organization."[28]

Thus, in the modern human brain the prefrontal cortex, as well as the frontal lobe as a whole, grows to a massive size. They appear to be no larger than one would expect given our body size and genetic similarity to the great apes, but such relative comparisons should not obscure their absolute magnitudes. The human brain ranges from 239 to 330 cubic centimeters in the volume of the frontal lobe as a whole, with the prefrontal area accounting for between 43 to 54 cm^3—these values dwarf those of a chimpanzee.[29] Furthermore, the human frontal lobe undergoes a remarkably long period of maturation that extends even into young adulthood. The synaptic pruning and myelination that occur over the first twenty or so years of life equip the human brain with a level and sophistication of executive control of working memory found in no other species. We human beings possess exceptional capacities for inhibiting impulses and delaying gratification as well as the ability to plan solutions to novel problems. Although human beings reveal individual differences in these abilities, the fact that we on average excel in such executive function must not be lost. Similarly, human failures to inhibit impulses, delay gratification, or solve novel problems are all too common. But such failings are often a source of regret and shame precisely for the reason that we recognize our inherent capacity to do better.

Cognitive neuroscience has identified a special network of executive attention that is mediated by structures in the frontal lobe.[30] It allows for the executive control of one's thoughts and behavior. The lateral prefrontal cortex, the subcortical structures of the basal ganglia, and the anterior region of the cingulate gyrus constitute the network for executive attention. This network allows the manipulation of information held temporarily in the storage areas of working memory. It is also used in the conscious effort to retrieve a fact or event from its residence in long-term memory and bring it into awareness for use in working memory. Although this network of attention is of particular importance in human cognition, the brain houses two other networks of attention that all mammals depend on for survival. One is the alerting network that ensures the brain remains in an awake, alert state of vigilance. The other is the orienting network that makes certain the head and eyes look in the direction of a novel visual stimulus and covertly focuses attention on a particular region of space even when the head and eyes are already in position. It is easy to see how the alerting and orienting networks were essential for any prey species that must avoid a predator.

At the heart of executive attention is an ability to deal with conflict. The brain is processing information in multiple systems all at the same time. To the extent that these systems lead to different needs, motivations, and responses at any given moment, there is a competition for the control of behavior. The anterior cingulate gyrus lies embedded within the interior of the frontal lobe. The cingulate gyrus extends the length of the brain from the frontal lobe to the parietal lobe, lying just above the white fiber tract known as the corpus callosum, which connects the left cortical hemisphere with the right cortical hemisphere. It is part of the limbic system that constitutes the forebrain of all mammals and is known to play an important role in emotion. In fact, the section of the anterior cingulate that is closest to the front of the brain, curling downward in the ventral direction, is closely linked with the amygdala. The amygdala works together with this emotional center of the anterior cingulate in the conscious awareness of fear.[31]

The other section of the anterior cingulate lies in a dorsal position just behind the emotional center, and it resolves cognitive conflicts. For example, if one habitually follows a certain route in driving home from work each evening, then the left and right turns can be made almost automatically. A driver can easily listen to the radio or daydream about her day or plans for the evening and the responses needed to get home are carried out effortlessly. However, suppose that before leaving from work the driver decides to stop at the grocery store to pick up something for dinner. This goal now sets up a cognitive conflict at that point in the route that calls for making a turn in the direction of the store instead of the habitual turn in the direction of home. Neuroimaging studies show activation of the cognitive center of the anterior cingulate gyrus in dealing with this response conflict. The automatic response must be inhibited and a response appropriate for the temporary goal must instead be selected. Cognitive neuroscientists study such conflict resolution with a laboratory task that simulates the same kind of situation. It is called the Stroop task. In its most common form, this requires reading the ink color of words that spell out primary colors, such as *red, blue, green, black,* and *yellow.* Word reading—at least in literate adults—is habitual and easily done automatically. A conflicting goal is established by deviously printing the color words in the wrong color ink—for example, *red* might appear in green ink. Because the goal is to read the ink color, one must inhibit the habitual response and select instead a response in conflict with it—say, green for the word *red* printed in green ink. When this task is done rapidly, it is very easy to make errors or to at least slow down considerably to make certain the correct response is made. The conflict situation not only slows responses and increases errors; it also requires greater effort to control behavior.[32]

Exactly what are the mechanisms that underlie the human capacity to control one's thoughts and behaviors? Three specific mechanisms have been identified that are carried out by the network of executive attention.[33] One is updating the contents of working memory by rapidly adding or deleting information depending on its immediate relevance. As new information bombards us from the environment or as new information is retrieved from long-term memory, the old contents of working memory must be discarded to make room for the new. Monitoring what should be discarded and what should be retained in the transient stores of working memory is thus a major part of successful updating. A second mechanism is shifting attention between two or more tasks or current goals. Because we often must keep in mind more than one goal or even attempt to do more than one task at a time, there must be an effective mechanism for shifting attention back and forth rapidly and flexibly. The third mechanism is inhibition. Through effortful control, inhibition allows the overriding of an automatic or dominant response in favor of one that is currently most appropriate. These are distinctly different mechanisms of cognitive control, but it is true that individuals who are good at one also tend to do well at the other two. Individuals who score well in updating, shifting, and inhibiting are more successful in regulating their thoughts and behaviors and resisting impulses and maladaptive urges. For example, they are more successful in "staying faithful to romantic partners . . . and successfully implementing dieting and exercising intentions."[34]

Although the network of executive attention takes a long time to develop fully, the third year of life brings with it major gains in updating, shifting, and inhibiting.[35] Toddlers at the age of twenty-four months are highly prone to perseveration—repeating a previous response, even though it is no longer appropriate. This tendency to keep repeating a behavior that is no longer appropriate is characteristic of adults who have suffered a brain injury to the frontal lobes. Both the toddlers and the frontal-lobe patients have difficulty with inhibiting a previous response, shifting attention to a new goal, and updating the contents of working memory appropriately.

Yet even after the third year it is apparent that executive control is still a work in progress. For example, the failure to inhibit inappropriate behavior is a hallmark of two-to-three-year-old children. The Simon Says game neatly shows just how difficult it is for such young children to inhibit a response until the brain has had nearly a full fourth year to develop. The game is played by performing an action when instructed by a toy bear but inhibiting the response when it is requested by a toy elephant. In fact, until the maturation of the third year of postnatal brain development is completed, the Simon Says game is simply beyond the executive attention capacity of the child.

The findings with Simon Says reveal that "children up to 40 months were unable to carry out the inhibition instruction at better than a chance level, and they performed just as rapidly in making incorrect responses to the elephant as they did in making correct responses to the bear. Although children could repeat the instruction, they did not seem able to use it to control their own behavior."[36] By the age of forty-eight months they performed it correctly almost every time.

An individual's ability to delay gratification is also related to differences in the functioning of executive attention. Waiting for a reward to come later requires the inhibition of taking immediate action to take the reward now. For example, consider a college student who is asked to go to a party early in the evening when he has plans to study for two hours before going. Resisting the impulse for the immediate social and emotional reward of joining friends at a party is necessary to obtain the larger reward of learning, completing assignments that will eventually need to be done, and preparing for an exam. The fun of the party will be added on top of these scholastic rewards. Human beings differ in their ability to delay immediate gratification, but the important point is that we, as a species, are capable of doing so. The default mode of the mammalian brain is to do whatever results in an immediate reward.

Delaying the impulse for immediate gratification requires an effortful inhibition of these urges, thus draining some of the limited supply of executive attention. If given the choice of a small immediate reward versus a much larger reward later, fully mature human beings have the capacity to delay gratification and gain the advantage of the larger reward. The prefrontal lobes and executive capacity develop only slowly during childhood and adolescence. It should not surprise us, then, that adults often make different decisions about taking risks to seek gratification than they did when in their youth. Further, those who develop strong executive control early in life tend to be more successful later in life. For example, preschoolers have been assessed in their ability to wait for a treat of two marshmallows that was superior to the one marshmallow available to eat immediately. Strikingly, the number of seconds that these young children were able to resist the impulse of immediate gratification predicted their ability to concentrate as an adolescent. It also predicted their ability as adolescents to deal with frustration and temptation.

The longer children could wait for the larger reward as a preschooler, the more likely they were to be judged as both academically and socially competent as adolescents.[37]

Delaying gratification is a very simple example of planning ahead or thinking strategically. More complex planning is essential to all problem solving. From where one now stands a series of steps must be taken that lead to a goal. Planning allows the problem solver to think one step or perhaps several steps ahead. It allows the problem solver to simulate in working memory—by talking about the plan covertly through inner speech or by imagining scenes in a visual-spatial format—all of the things that could go wrong when a step is taken in a particular direction. The human ability to hold different scenarios in mind and to try out different possible paths to the goal thus depends on executive attention.

Although human beings are superior in executive control compared with other species, it must be remembered that the resource of executive attention is limited in us all. It can readily be fatigued by constant demands, for example. Individuals with lower capacities of executive attention have more difficulties in coping with the incessant need for self-control of impulses. Moreover, the larger the capacity to attend to mental representations and maintain them in working memory, the better one can comprehend speech, read, write, reason, solve problems, and think in general. These mental skills are collectively assessed by a test of general fluid intelligence. Scores on such tests can be accurately predicted by knowing the capacity of executive attention.

Cognitive psychologists have devised a test to measure individual differences in how well one can update the contents of working memory, shift between two tasks, and inhibit old irrelevant information. The test involves two tasks that have to be performed concurrently. A mental arithmetic problem is presented first, followed by a word. The test takers calculate the answer to the arithmetic problem and then shift attention to storing the word in working memory. Then another arithmetic problem and word are presented. After several of these trials, the person must try to recall all of the words that had been presented up to that point. Individuals with a large capacity for storing the words in verbal working memory, and, most importantly, a large capacity for executive control, can remember the most words correctly after correctly performing each arithmetic problem. Scores on this test of executive attention and working memory storage correlate moderately to strongly with general fluid intelligence.[38]

Fluid intelligence refers to the ability to solve novel problems. It differs from the breadth and depth of knowledge that a person has learned and stored in long-term memory—a construct known as crystallized intelligence. Although both forms of intelligence are important for human beings, the capacity to solve novel problems provides the engine of innovation that propels cultural evolution. It is notable, then, that the capacity of executive attention is such an important determinant of fluid intelligence.

Neuroimaging of the human brain while it solves novel problems taken from standardized tests of fluid intelligence has revealed portions of the network of executive attention. For both verbal and spatial novel problems that require a high degree of fluid intelligence, the lateral region of the prefrontal cortex in the left hemisphere showed strong activation.[39]

Thus, the massive frontal lobes of the human brain provide an abundance of cortical space for the support of an advanced system of working memory and, in particular, its central executive component. The extended juvenile period that characterizes human development provides an extended period for pruning synapses in the frontal lobe of the brain and for completing the myelination of neural axons that allows for highly efficient neural networks. Metaphorically speaking, the executive suite of the human brain is well furnished by the time it takes one to reach young adulthood. The adult human brain allows a level of executive control that is never attained at all in other primate species. In commenting on comparative research with mature rhesus monkeys, for example, Michael Posner and Mary Rothbart observed that the "results of the tests with monkeys seemed more like those of young children whose executive attention is still immature than like those of adult humans."[40] Human beings eventually develop a capacity for delaying gratification, inhibiting inappropriate behaviors, and planning solutions to complex novel problems.

The human capacity for fluid intelligence—the ability to solve novel problems—is of particular importance in understanding the unique cultural world in which we live. Human beings stand apart from other species in our capacities for behavioral innovation and the cultural transmission of those innovations from one generation to the next. It has been argued in the present chapter that an advanced form of working memory is critical for the innovation part of this equation. In the next chapter, the advanced form of social intelligence found in human beings will be addressed. Although human beings have attained a highly advanced form of both working memory and social intelligence, there is nonetheless evidence of continuity with other primates with respect to both. Instances of behavioral innovation have been observed in numerous primate species—this refers to behaviors observed in the field or in captivity that were never seen before, as opposed to traditional behavior patterns of the species. Examples of social learning have been documented as well. By tallying the frequency of such instances of innovation and social learning across different species of primates, it was possible to examine the relation of both to brain development.[41] The volume of the neocortex and striatum taken together provided an index of what the researchers called the "executive brain." This reflects in part the sheer amount of cortical capacity available for executive attention and the other components of working memory in nonhuman species. Because primate species differ widely in body and overall brain size, the executive brain ratio was computed by dividing the neocortex and striatum volume by the volume of the brainstem. In essence, the executive brain ratio calibrates the extent to which the forebrain can support an advanced working memory and advanced social intelligence. The researchers found a significant positive relationship between the executive brain ratio and both innovation and social learning. In fact, innovation and social learning were also correlated with each other, indicating that the two are not entirely independent of each other. Thus, even while confined to nonhuman primates, there is a clear relation between the amount of cortex that supports the executive functions of working memory and the likelihood of engaging in novel behaviors or learning behaviors from observing others.

In sum, the executive function of human working memory is highly advanced, and it undoubtedly played an important part in the emergence of the modern human mind. Another major change is the addition of a phonological loop specialized for the learning of language. Yet continuity with nonhuman species is also not difficult to discern. Primates in general, including even the relatively small-brained rhesus monkey, make use of the prefrontal cortex for the temporary storage of stimuli that are currently out of sight. Baboons are severely limited to storing only about four chunks of information in the visuo-spatial sketchpad of working memory, but then so are human beings. Although human beings possess a much larger frontal lobe than other species, it is not larger than what would be expected for a great ape with a similar overall brain size. Thus, it is not relative space allocation but rather the connectivity of the brain's neural circuitry that appears to be where a reorganization of the frontal lobe took place in the transition from nonhuman to human. The puzzle piece of working memory appears different in human beings than in our close genetic neighbor, the chimpanzee, but not unrecognizably so. It is when the piece is put together with other parts of the modern ensemble that a qualitatively different kind of mind emerges. One example of this interaction has already been described with the addition of the phonological loop in human beings as an adaptation for the learning and use of language.

Chapter 3

SOCIAL INTELLIGENCE

The second part of the ensemble of the modern mind is an advanced social intelligence that enables human beings to invent—through active collaboration—the language, beliefs, behavioral practices, and societal traditions that envelop us in our social environment. The innovations made possible by an advanced form of working memory alter the natural world into a social and technological environment of our own design. The cultural world that we invent serves as communal glue that binds us together with common values and ways of living. It invades the brain from the moment of birth, creating a mind shared by many. It is information as valuable to pass on to the next generation as the genetic information encoded in DNA. Just as DNA transmits the traits of our parents and direct ancestors forward to us, so, too, does the culture of "our people" replicate and survive into the future. The advanced social intelligence of human beings provides us with the tools to observe others and imitate their behavior. Through these tools, the mantle of culture is passed from one generation to the next.

Besides providing the means for the inheritance of culture, our advanced social intelligence enhances our capacity to innovate new ways of living. It opens us to the possibility of collaboration among individuals for mutual gain, rather than relying exclusively on competition for self-survival. In *Homo sapiens*, individuals both learn from each other and work together to solve common problems. Through such collaboration, the power of human fluid intelligence to innovate is leveraged many times over. The executive functions of working memory provide the ability for any human being to solve a novel problem. But imagine multiplying that ability by five, ten, or twenty times as groups of human beings collaborate together. Here, then, is a prime example of an interaction between two parts of the modern mental ensemble. From this interaction, innovations flourished and were socially transmitted from one group of human beings to another, and especially from one generation to the next. Rapid innovation plus cultural inheritance released human beings from the glacially slow pace of biological evolution. Tens of thousands, if not hundreds of thousands, of years were no longer needed for dramatic changes in the functioning of the human mind. Through an advanced social intelligence and its leveraging of fluid intelligence, human beings were launched onto the fast track of evolution—the evolution of culture rather than genes.

Culture—the words used to refer to ideas and events, the political and religious beliefs, the preferences for clothing and food, and thousands of other practices and traditions—sets us in a different world from all other species. We simultaneously inhabit a physical world—one we certainly share with chimpanzees—and a cultural world, which is reserved for the human mind. Human beings swim in culture in much the way fish swim in water. Culture surrounds us, and we breathe it in as a kind of mental oxygen. It is the sum total of the human way of life, accumulated through history and inherited from generation to generation since the Stone Age. People living in close proximity in a group often develop common beliefs about the world, just as they come to share food preferences, clothing styles, means of adorning the body, tools, and language. The group finds common ground for social interactions among its members and expectations of what is normal and what is deviant. Such norms provide the social glue that holds the group of people together.

Because culture is everywhere in our lives, it is generally invisible to us, so taken for granted as to be imperceptible. But such transparency can quickly turn opaque when human beings migrate from one part of the world to another. Imagine an American raised on a farm in rural Kansas transplanted to the Asian metropolis of Shanghai. Or imagine the life-long New Yorker relocated to the remote mountains of Afghanistan. For immigrants, the pain of separating from family, friends, and the country of one's childhood is compounded by the uncertainty and novelty of the adopted land. *Culture shock* aptly captures the disorientation of being dumped from familiar to foreign cultural waters overnight. The immigrant must acclimate to the strange waters and learn to swim again in the new culture. Travelers, too, experience culture shock, although only temporarily while visiting a foreign country before returning to the familiarity of home.

Human beings are a richly diverse species. The differences among human populations, however, are much less a matter of genetic differences and much more a matter of cultural differences. The sounds we utter, the clothes we wear (or, equally significant, do not wear), the food we eat (and the way we eat it), the rules we follow in living together, and the beliefs we hold about government, health, ethics, religion, destiny, and all else in between separate us from one another in ways that our genes do not.

A definitive feature of the human mind is our massive addition to the physical world of things that we need or imagine we need. For example, the human capacity to make tools, from prehistoric stone tools to the machines of the twenty-first century, has transformed the physical landscape. The invention of agriculture prior to 4000 BCE laid the foundation. The invention of industry in the late nineteenth century yielded a massive expansion of artifacts to satisfy the material, psychological, and spiritual needs of human beings. The globalization of industry, driven by technological innovation, in the late twentieth century vastly accelerated the process. Buildings, furniture, appliances, books, blogs, paintings, photographs, sculptures, graffiti, musical scores, telephones, cell phones, Blackberries, iPods, iPhones, desktop computers, laptop computers, and plastic containers in every conceivable shape and size, are now as much a part of the biosphere as plants and animals.

THE FAST LANE OF EVOLUTION

The cultural practices of modern human beings in the Upper Paleolithic or Late Stone Age of thirty-five thousand to eight thousand years ago have been deciphered by archeologists using evidence from human artifacts dated to the time period. As discussed in chapter 1, the cave paintings and engravings of the Cro-Magnon people of southern France provide an example of such evidence, and similar images of rock art have been found in nearby Africa and in distant Australia. Such art first emerged during the Upper Paleolithic, and it marked the advent of the modern human mind. The number, variety, and sophistication of tools, weapons, and bodily adornment also exploded.[1] Most significant for the point made here is that the rate of cultural change began to accelerate during the Upper Paleolithic, and it accelerated still more in the Neolithic period, continued unabated into the ancient world, and today overwhelms us with scientific and technological innovation almost on a yearly basis.

Early modern humans lived as roving bands of hunters and gathers, a means of survival that dates back millions of years to the origins of primates. However, by eight thousand to five thousand years ago, this nomadic lifestyle of tracking down the food sources needed for survival changed momentously during the Neolithic into a lifestyle in which small groups of villagers farmed resources. The invention and evolution of agriculture occurred rapidly compared with the long static period of hunting and gathering practiced by the immediate and deeply remote ancestors of *Homo sapiens*. Then, in just a few thousand years, farming villages expanded and became more socially and politically complex, with larger and larger populations, leading to temple towns, city-states, and eventually national states.[2] By five thousand to three thousand years ago, state societies had appeared with a specialized institution of government. Having evolved from small groups of hunters and gatherers with relatively simple social organizations, state societies became richly diverse and complex, with individual members differing widely in wealth, status, and political power to influence others. Consider ancient Egypt, for example. The pharaoh governed all and commandeered the resources of the professional classes, such as engineers and accountants, as well as the physical labor of slaves to erect his own burial tomb for the afterlife. The shift from hunter-gatherers to the complex society of Egypt took, of course, multiple generations. Yet all this happened in the blink of an eye when framed against the time scale of millions of years of biological evolution and the deep time of geology.

The invention of pottery illustrates that innovation began to pick up steam during the Upper Paleolithic.[3] Firing ceramic containers is a sophisticated step ahead of baking small clay objects and figurines in terms of its technical demands. The invention of pottery also implied an interest in cooking and preparing and storing food. Archeologists traditionally posited that first came the invention of farming and permanent settlements—an advance over the hunting-gatherer culture that endured for tens of thousands of years—followed by the invention of pottery. It was believed that agriculture and pottery were thus innovations of the Neolithic period and were certainly no older than eight thousand years. The recent discovery of shards of pottery from the Xianrendong Cave in the Jiangxi Provence of China convincingly falsifies this assumption. Radiocarbon ages of the site indicate that pottery had been invented nineteen thousand to twenty thousand years ago in the Upper Paleolithic. The technology of pottery thus preceded agriculture. It was a much earlier innovation of hunter-gatherers to cook their food: "Pottery making introduces a fundamental shift in human dietary history, and Xianrendong demonstrates that hunter-gatherers in East Asia used pottery for some ten thousand years before they became sedentary or began cultivating plants."[4]

Cultural change happens rapidly, as the relatively rapid ascent from hunter-gatherers to nation-states illustrates. The acceleration of cultural change in modern humans has continued to the point that within a couple generations the world can change so as to be almost unrecognizable. Consider the technological innovations of the past hundred years. In an editorial on consumption as an indicator of economic well-being, the *New York Times* published in its Sunday Opinion section (February 10, 2008) a chart plotting the percentage of US households with specific technologies from 1900 to 2005.[5] For the argument advanced here, the chart (prepared by Nicholas Felton and titled "Consumption Spreads Faster Today) documents how quickly technological innovations penetrate our culture in recent years compared with a century ago. The telephone and electricity were scarcely used in 1900 and the automobile was as yet unknown. Radio and refrigerators were unknown until 1920. Air conditioning, clothes dryers, and dishwashers did not emerge until nearly 1950. Yet all of these technologies are taken for granted by Baby Boomers born after World War II. The acceleration of cultural change in the first half of the twentieth century altered the technological world of a Baby Boomer's parents and grandparents profoundly. As stunning, if not more so, is the explosion of technology during in the second half of the twentieth century. Among the technologies taken for granted by the Boomers' children are microwave ovens, personal computers, cell phones, smart phones, and the Internet. It took less than twenty years for microwave ovens to penetrate into

80 percent of all US households. Telephones and automobiles required sixty years to do the same! It took more than forty-five years before 90 percent of households had electricity. From 1990 to 2005, cell phone use jumped from less than 5 percent of households to 90 percent, while personal computers did much the same by entering more than 70 percent of all households from less than 20 percent in the same fifteen-year interval. Only radio and refrigerators from the first half of the twentieth century showed adoption rates even close to those of the technological innovations of its last quarter.

Human beings alone manufactured advanced stone tools and symbolic art, invented and propagated language, and organized small villages and eventually large nation-states. These astounding cultural advances have taken place in a geological blink of the eye since the origin of the modern human mind a few tens of thousands of years ago. As Michael Tomasello has argued, biological evolution could not possibly explain the cultural world of modern humans. Our genetic similarity to other great apes is "the same degree of relatedness as that of other sister genera such as lions and tigers, horses and zebras, and rats and mice. . . . There simply has not been enough time for normal processes of biological evolution involving generic variation and natural selection to have created, one by one, each of the cognitive skills necessary for modern humans to invent and maintain complex tool-use industries and technologies, complex forms of symbolic communication and representation, and complex social organizations and institutions."[6]

The radically different process of change underlying the human story is called cultural evolution. It alone can account for the acceleration of change that has characterized the history of our species. Cultural innovations added in one generation are carried forward into the future and are generally not forgotten and lost. Rather, they are learned by the next generation, either through imitation or by direct instruction. This allows innovations to accumulate from one generation to the next. The accumulation of past cultural accomplishments explains why the pace of change accelerates through time. As the cumulative total of innovations grows larger and larger, there are more and more places in which to innovate further. Thus, one would expect an acceleration of change from prehistory to modernity. Now, in the twenty-first century, after tens of thousands of years of cultural innovation, one should not be surprised to see dramatic changes over the course of as little as one hundred years.

The *ratchet effect* refers to the accumulation of innovations and modifications over multiple generations.[7] Cultural traditions have the possibility of becoming more rich and complex as a result. Children in each generation learn from their parents about the traditions of the past. This cultural learning process serves as a ratchet to prevent the culture from slipping backward and losing the innovations of earlier generations. Of course, some insights, inventions, and ways of living are in fact lost over time, but the ratchet effect of children's cultural learning can succeed on the whole, at least for those cultures that show cumulative traditions. One clear example from prehistory to the present is the evolution of hammers. The design of hammers has changed repeatedly over thousands of years, "going from simple stones, to composite tools composed of a stone tied to a stick, to various types of modern metal hammers and even mechanical hammers."[8]

The ratchet effect that accounts for the cumulative nature of cultural evolution thus depends on both innovation and learning. In the ratchet effect, one can see how an advanced social intelligence exploited the innovative capacities of an advanced system of working memory. Innovations arise from the fluid intelligence made possible by the executive attention of human working memory. Children can then learn of past innovations from their parents and other elders in their community through a process of imitation and practice, or at times through direct instruction. Imitative and instructed learning serve as ratchets that hold fast the innovations of the previous generation. Through them, these innovations can be faithfully brought into working memory among members of the current generation. This allows further discoveries and further modifications that can then be passed on to the next generation. Without the ratchet of cultural learning, the discoveries of human beings would simply die with them. Instead, creative individuals living now can in a sense collaborate with inventors who came before by taking up where they left off to innovate further.

If one thinks of biological evolution as the slow, measured way of creating life in general and the human brain in particular, then cultural evolution was the move to the fast lane. The modern human mind benefited from an entirely new means of evolving through inheritance of past adaptations. This new means of cultural inheritance rushed like a mountain stream past the glacial pace of biological inheritance.

CULTURE MOLDS THE MIND

Whereas human populations are largely identical to one another at the level of the human genome, cultural differences in thought, values, beliefs, and behaviors are enormous. Ethnic differences are so potent that they often blind us to our common humanity. Although all human beings are born with the same nervous system and brain, the culture in which an individual is raised molds the mind from birth and throughout the lifespan. Although we come into the world equipped to learn the sounds used in any human language, the mind becomes attuned to hearing and producing only those phonological segments that constitute the phonemes of one's native language. Although all children are capable of using nonverbal gestures and expressions to communicate, cultures differ in the extent of use and degree of animation of body language. In a similar way, the acceptable physical proximity while conversing and the expectation of social hugs and kisses while greeting vary with culture. Or, consider the variations in what people find appetizing to eat for dinner. An American typically eats pork and beef, whereas an Israelis might shun pork or an Indian might shun beef on religious grounds. Americans and Europeans are squeamish about snake, whereas the Chinese enjoy it as well as pork and beef. Yet Europeans may enjoy eels and snails, neither of which is marketed to Americans, despite their tradition for consuming raw oysters. None of these ethnic variations can be accounted for in the biology of the brain or stomach—they reflect instead the shaping of the human mind by the social world into which we are born.

Psychologists have documented culturally induced differences in basic mental processes by comparing one demographic or ethnic group to another. Much of this research has contrasted East Asians with European North Americans to specify these differences at a detailed level.[9] Americans who are descendants of immigrants from Europe belong to a culture that stresses individualism, self-reliance, and independence. By contrast, East Asian cultures stress collectivist ways of thinking and the interdependence of group members. Collectivist thinking places the needs of the group as a whole above the needs of the individual. These culturally mediated differences in how the mind functions permeate daily life. When interpersonal conflicts arise, Americans often adopt confrontational approaches, including even filing a lawsuit—the sledgehammer of direct confrontation. East Asians, by contrast, seek mediation and accommodation as a way to resolve conflicts with a minimum of animosity among the parties involved. Marketing strategies differ, with advertisements appealing to harmony and benefits for the group as a whole in East Asia while ads for Americans appeal to a desire for competitive advantage and individual uniqueness. American mothers expect greater independence in their children at a young age compared with East Asian mothers. In school, the East Asian teacher is more likely to address communication to the class as a whole whereas American teachers tend to communicate directly to the individual student.

As part of normal cognitive development, the human mind constructs a concept of the self during childhood and adolescence. Relative to Americans, when young adults in East Asia are asked to describe the self, they generally make more statements about the groups to which they belong and the ways in which the self is interdependent on others. East Asian culture "prescribes devaluation of one's distinctive personal strengths that are unrelated to or would even hinder actualization of collective goals . . . and a strong motivation to avoid failures that would reflect badly on the group."[10] By contrast, the independence stressed by Western culture biases Americans toward "self-enhancement (viewing one's personal attributes as better than they really are), unrealistic optimism (perceiving the self as more invulnerable and more likely to experience positive events than it really is), and self-affirmation (justifying one's personal choices)."[11]

The power of culture to mold the mind can also be observed in those who migrate to a new culture. Although one's native culture is never left behind completely, immigrants must undergo a process of acculturation and change in response to their exposure to a new language, new ways of living, and new values and beliefs. The nature of acculturation varies markedly with the age of the immigrant.[12] Young children are far more flexible in identifying with the receiving culture and learning its practices, including its language. Migrating as an adolescent or an adult implies leaving behind a culture that has already profoundly sculpted the individual's mind. This is especially true of older adults for whom it is easier to retain the language and identify of the heritage culture instead of attempting to acquire the new one. Ethnic enclaves within the receiving culture offer a buffer and allow retention of the heritage culture in large measure. For example, in the United States, parts of Miami, the South Bronx, and Chinatown neighborhoods in various cities offer protective ethnic enclaves. The heritage culture within them is so strong that even the second-generation immigrants—those born into American culture—can be influenced to retain the culture of their parents.

The advanced social intelligence of human beings provides the obvious explanation for why one's native culture imprints on the mind. The human brain is designed to attend to the behaviors and minds of others; it is designed to learn by observation and imitation. All this is made easier by the highly immature state of the body and, in particular, the organ of the brain at birth. Unlike many other mammals and primates, human beings are altricial and completely dependent on parental care during infancy. Unlike, say, the prococial horse, ready to stand at birth, the altricial human being is utterly helpless. As can be seen in the following facts, the immaturity of the brain at birth is especially significant for the impact that culture has on the developing mind.[13] Whereas the sensory areas mature early in development, the higher cortical areas mature later, after exposure to environmental influences. The brain is still growing throughout childhood, when the individual is largely dependent on and immersed in the cultural environment of the parents; the brain does not reach adult size until around puberty. Finally, the prefrontal cortex is not fully mature until young adulthood, after two decades of cultural influence.

Culture can thus imprint upon the developing human brain and sculpt an individual's mind to fit into an ethnic group. The brain—in an immature, still developing state—is bathed in the physical and social environment of the child's immediate world throughout infancy, childhood, adolescence, and even young adulthood. Just as the fetal environment of the mother's womb bathed the brain during prenatal development, so, too, does the cultural world into which the child is born. Nature has left much of human brain development for later, after birth, when nurture can shape the outcome rather than leaving it to a genetically predetermined plan. The language, social norms, and ways of life of the child's native culture are thus melded into the neural circuitry. Because human brain maturation is delayed, the advanced social intelligence that we have inherited in our brain structure and functioning produces a mind determined as much by culture as by DNA.

ADAPTATIONS FOR CULTURAL LEARNING

By the age of nine to twelve months, infants are normally able to share eye contact with another person.[14] Eye contact between the mother, father, or other caregiver and the infant becomes a medium for sharing the same focus of attention. For example, if a parent looks at and points to a cup on the tray of a daughter's high chair and the infant directs her gaze at the cup at the same time, the parent and child are in essence joining minds for a moment. Both are perceiving the cup and holding it for a while in working memory, allowing the possibility of the infant learning something new from the parent. The parent might say "cup" or "this is a cup" or "do you want your cup." In each case, there is an opportunity for the child to learn the name of the object held now in the focus of attention.

Joint attention narrows to one the number of possible objects in the room that *cup* could refer to, without which the child would be lost trying to figure out whether *cup* refers to the tray, the floor, the spoon next to the cup, and so on. Learning the names of objects should thus be greatly aided by joint attention. In fact, the ability of infants to follow the gaze of an adult at the age of ten to eleven months is predictive of fluency with language at eighteen months of age.[15] This, then, is one illustration of how the infant's response to the caregiver's focus of attention is the basis for learning culture through social interaction with another, more knowledgeable human being.

Infants also initiate joint attention to share with others what interests them. A child might point to an object in order to attract the parent's attention to it. Or the child might gaze at the object and then look into the parent's eyes and then look back at the object. The child might repeat this several times in order to establish joint attention.

Developmental psychologists have learned that such initiation of joint attention is surprisingly not closely connected with simply responding to the gaze and gestures of the caregiver. The two ways for minds to meet are mediated by separate systems of attention in the neocortex.[16] Responding to joint attention is mediated by the parietal and temporal lobes in the posterior regions of the brain. The frontal lobe of the brain provides the neural networks for an anterior attention system that can initiate gestures and control gaze direction. It is part of the executive attention component of working memory, and it develops at a slower pace relative to the posterior attention system.

Recognition of Intentional Agents

Importantly, joint attention requires the capacity to hold in mind a triadic relationship of the child, the adult, and the object or event being attended to by both. By six months of age, infants can interact in a dyad with an object or with another person. Grasping a cup, for example, involves only the child and the object. Or making eye contact with the adult involves only the child and adult. But in joint attention, all three elements of the triangle are represented; this triadic capability emerges later in cognitive development, between nine to twelve months. Within this four-month window, almost 80 percent of infants show numerous signs of skill in joint attention.[17] Not only gaze following, but also point following, imitation of acts, and the use of gestures to make declarations of facts and imperative demands all coalesce during this interval.

Why does joint attention begin to emerge at nine months rather than earlier? Michael Tomasello argues that "infants begin to engage in joint attentional interactions when they begin to understand other persons as intentional agents like the self," as "animate beings who have goals and who make active choices among behavioral means for attaining these goals, including active choices in what to pay attention to in pursuing those goals."[18] The infant needs to grasp that the adult has goals, just like he or she does, and that the adult can make choices about how to attain these goals—in other words, the adult has a mind just as a child has a mind. As cognitive development proceeds, infants and young children develop increasingly sophisticated understandings about the minds of others. The process is only just beginning at the end of the first year of life.

Children can be thought of as developing a theory of mind as they come to understand others as intentional agents with minds of their own.[19] That is to say, the child understands that other human beings have beliefs, desires, and thoughts about how to get what they want. Initially, the infant is like the solipsistic philosopher who argues that only he is exists as a conscious being, because the self has immediate access only to its own mind. As the infant matures and learns about the way animate agents move and act in the world to obtain things, she comes to see that others have minds as well. The social nature of human beings includes the capacity to engage in a form of mind reading—that is to say, the ability to use clues from eye gaze, facial expression, and other indicators of thought and attention to imagine what another person is thinking or believing.

By the end of the second year of life, young children are building up their ability to read the minds of other people, but this development slowly unfolds. Between the age of two and four, a theory of mind is still a work in progress, and the child is unable to grasp explicitly that his or her own belief can be different from the belief of another person. Yet, by about the age of four, the preschool child will have developed a relatively advanced theory of mind characteristic of adults—the four year old will understand that the beliefs, desires, and thoughts of other people may not be the same as their own. This transition has been measured using a test of false beliefs. For example, suppose that a young child observes the following scene. Sally enters the room where Anne is sitting and hides a marble in a basket. Sally then leaves the room, giving Anne a chance to trick Sally by getting up, retrieving the marble, and moving it to a box. Finally, Sally re-enters the room to get the marble. Where will she look for it? For three-year-old toddlers, without an advanced theory of mind, they respond that Sally will look in the box. They do not seem to be able to grasp that Sally holds in mind a mistaken belief about where the marble is, namely, in the basket where she left it. The three year olds assume that Sally's understanding of the situation is identical to their understanding. Yet, within another year, by the age of four, children readily grasp the notion that two people can hold different beliefs and that another person can hold a false belief. The four year olds respond that Sally will look for the marble in the basket.[20] Between the age of three and four, the typical progression of human brain development allows the child to read the mind of Sally, in a sense, and grasp that her perspective is distinct from the child's own.

Can the typical developmental progression of coming to recognize intentional agents go awry? It seems to in the developmental disorder known as autism. Autistic children can in some cases have normal levels of intelligence. In fact, autistic savants show extraordinary cognitive abilities in specific domains, such as artistic or mathematical abilities. Yet autistic children have difficulty seeing other people as intentional agents with minds of their own.[21] They are in a sense blinded to a degree to the minds of others, a social form of blindness that disrupts interactions with other people and efforts to communicate. Autistic children instead risk becoming isolated in a world of their own making rather than fully experiencing the shared cultural world.

Imitation

Joint attention and the recognition of intentional agents are important milestones in the first important form of cultural learning. Without these two steps, the next step of social learning through the imitation of others cannot be taken. At around nine months of age, the infant begins to reproduce the intentional behavior of an adult.[22] For example, if the caregiver models drinking from a cup, then the infant will try to imitate the behavior. The infant adopts the goal of wanting to get a drink and follows the behavioral lead of the adult in how to achieve it.

Less complicated kinds of imitation start practically at birth in face-to-face dyadic interactions between the infant and parent.[23] If a father sticks out his tongue at his newborn son, the baby can stick out his tongue to mimic the parent's facial gesture. This shows that social learning begins at a very early age, preparing the infant for the more complex triadic imitation that emerges several months later. The nine-to-twelve-month-old-infant can recognize the parent as an intentional agent and can hold in mind the adult and an external object and the self in a triad of joint attention. During early childhood, all sorts of parental behaviors are imitated: "For example, a toddler may see her father using a telephone or computer keyboard and crawl up on the chair and babble into the receiver or poke the keys."[24]

Cultural learning—observing and imitating the intentional actions or behavior strategies of others—provides a way for one generation to pass on its knowledge to the next generation. Learning in this way is certainly not limited to young children. For example, apprentices in crafts and trades learn skills through imitation and practice as well. Although imitation is an important form of cultural learning, it is not the only means by which knowledge can be faithfully passed from one generation to the next. A teacher can also directly instruct the learner in how to do something. Direct instruction forms the core of schooling as a way of propagating cultural practices through historical time. Another form of cultural learning depends on the invention of written language. Once the knowledge of a teacher has been written down, the learner can read the information without the teacher needing to be present. Instruction can thus also occur through reading, at least among literate learners.

DEEP ENCULTURATION
AS UNIQUELY HUMAN

The cognitive development of other species, like us, depends on genes passed down from ancestors and on the influence of the current environment on the expression of the genes in the brain and body, what biologists call epigenetic factors. However, in human beings alone there is a deep or cognitive enculturation that structures the brain and mind in profound ways not seen in any other species. Parents, family, schools, church, and other institutions of society assume control of the child's cognitive development shortly after birth. The brain adapts to a culture filled with symbolic as well as material artifacts. For example, literacy brain circuits are developed for reading and writing as a consequence of children growing up in a social world of communication and linguistic symbols. Mathematical circuits emerge through schooling; these take the core concept of number prewired in the brain through genetic coding and grow a complex system of mathematical literacy. Musical literacy also alters the brain's architecture through immersion in a family and community of singing, dancing, and instruction in the musical arts.

Chimpanzee Traditions

Biologists have looked for behavioral traditions in different communities of wild chimpanzees. If any other species were to show signs of cultural evolution, one might expect something similar in our closest genetic relative. Research has shown that different groups maintain as many as thirty-nine different traditions.[25] Other mammals, birds, and fish, far more distantly related to us, typically display only a single behavioral tradition and never more than a handful. To illustrate a chimpanzee tradition, consider the style of clasping hands used to initiate grooming. Whereas one group of chimpanzees clasped hands palm to palm, another group in a nearby region used a wrist-to-wrist style. As another example, the ways chimpanzees scratch each other's backs also varies by community, some using a long, raking scratch and other groups preferring a short jabbing technique.

Most strikingly, a community of chimpanzees in the Republic of the Congo invented a pair of tools to fish for termites for food in their mounded nests.[26] One tool is a stout stick that can be jammed down into the ground to expose a tunnel into the termite nest. Another tool is a slender probe that the chimpanzee first prepares by stripping its leaves, shortening it to a good length by biting, and then pulling it through the teeth to fashion a nice brush tip. This tool is then inserted into the tunnel to collect and extract the termites, like a fish on a hook. These termite-fishing methods have not been observed beyond this community. By contrast, some populations in Western Africa simply smash open the termite mound with large sticks and then grasp the insects with their hands.

Although chimpanzees have many more behavioral traditions than other species, they do not display a cumulative culture the way human beings do. For chimpanzees, there is little if any indication of innovation and improvement as a tradition is passed from one generation to the next. By contrast, we invent a material artifact, such as a tool, and then pass it down to the next generation, where it might be either improved upon or at least transmitted faithfully and not simply lost in the rubble of history. Through the ratchet effect, human culture progresses by building on the wisdom of the past. Human beings not only improve the sophistication of their material tools, but also do the same with symbolic artifacts of, say, language and art.

The cultural gulf between chimpanzees and humans is understandable for at least two reasons. First, the behavioral traditions and tools invented by chimpanzees are relatively simple and not readily modifiable into more complex, improved versions. They may lack the cognitive capacity needed to invent a more complex tradition or tool. Whereas human beings are endowed with a strong fluid intelligence that stems from our advanced working memory system, innovation is not second nature for chimpanzees. Second, the chimpanzee is not as adept at cultural learning through imitation as a human being is. The advanced social intelligence of human beings specifically equips us for cumulative culture in a way not seen in chimpanzees. This is not to say that apes cannot ape. But they are more likely to engage in a form of social learning known as emulation rather than imitation. By emulating the behavior of a model, one learns that the model did something to change the state of the environment. But the change in the environment is the focus of the learning rather than the specific behavioral strategy used by the model.

Emulation versus Imitation

Michael Tomasello, in *The Cultural Origins of Human Cognition*, observed that chimpanzees are very good at learning how to obtain goals by observing the behavior of their conspecifics.

> For example, if a mother rolls a log and eats the insects underneath, her child will very likely follow suit. This is simply because the child learned from the mother's act that there are insects under the log—a fact she did not know and very likely would not have discovered on her own. But she did not learn from her mother how to roll a log or to eat insects; these are things she already knew how to do or could learn how to do on her own. (Thus, the youngster would have learned the same thing if the wind, rather than her mother, had caused the log to roll over and expose the ants.)[27]

Tomasello went on to describe an experiment that he and his colleagues conducted to contrast the emulation of chimpanzees with the imitation of human beings.[28] They gave a rake-like tool to a chimpanzee, along with an object that was just out of reach. The rake was designed so it could be used either in an efficient way to retrieve the object or in an inefficient way. The experiment was designed so that one group of chimpanzees saw a model that demonstrated how to use the rake efficiently while another group observed the inefficient method. The results showed that the chimpanzees used the rakes in all sorts of ways after watching the demonstration, and it made no difference at all whether they had seen the efficient or the inefficient demonstration. Thus, they learned through emulation that the rake could be used in some fashion to obtain the object, but they did not imitate the behavioral strategy for doing so.

The experiment was repeated with two-year-old children with a fascinating difference in outcome. The human children generally copied the method of using the rake that they saw in the demonstration assigned to them. This proves that they were imitating the strategy, not just emulating its effect on the environment. Most importantly, the children imitated the inefficient method just as faithfully as the efficient one! As a consequence, the children who saw the model use the rake in an inefficient way ended up with a poorer level of success in retrieving the object compared with the chimpanzees. The inclinations of human beings to imitate is so powerful that it can stand in the way of finding creative ways to solve problems that do not rely on reproducing the ways observed in the past. This is intriguing, for it shows that "imitative learning is thus not a 'higher' or 'more intelligent' learning strategy than emulation learning; it is simply a more social strategy."[29]

Other experiments have documented the same conclusion; human beings are so attuned to the behavior of models that they imitate them even when it makes little sense to do so.[30] Imagine a box with a desirable treat on the bottom and two openings, one on the side and one at the top. Further imagine that the top opening leads only to a platform in the middle of the box that prevents probing with a tool to reach the food on the bottom. A human model demonstrated how to get to the food by first trying the top opening and then moving the tool into the side opening to obtain the treat. When the box is covered so that the test subject cannot see inside, both chimpanzees and human beings followed the lead of the model—they both imitated what they saw the model do. The interesting question was what would happen when the box was made transparent, when it was possible to see that a platform in the middle prevented any success with using the tool in the top opening.

The results with this transparent box showed that "the chimpanzees switched from the imitative approach they had used with the opaque box to a relatively emulative strategy that focused on the crucial terminal work around the low hole."[31] Stunningly, the children continued to conform to what the model had done—they ignored what their own eyes told them and imitated the model!

What seems to prevent chimpanzees from relying on imitation as a primary means of social learning could well be their inability to recognize other chimpanzees as intentional agents.[32] The goal or intention of a model is readily picked up by a human being, as is the specific means used to achieve the goal. The means and the end are separable in the human's mind. By contrast, the intentions of the model and the behavioral means used to achieve the goal are probably not part of the chimpanzee's experience in witnessing the whole event. Instead, the chimpanzee possibly only perceives changes in the environment, and the demonstrator's actions are just part of these changes. Unlike human beings, chimpanzees seem unable to perceive these environmental changes as parts of a behavioral strategy that can be copied.

Although human beings possess an advanced form of social intelligence, efforts to enculturate apes by training them in imitative learning have shown some success. Even though wild chimpanzees do not show signs of the capacity for joint attention, there is sufficient continuity between the two species for intensive training to yield some positive results. For example, wild apes do not appear to point to an object for others to look at or to hold up an object to show others. Yet these skills can be trained to a degree, most likely because "in a human-like cultural environment . . . they are constantly interacting with humans who show them things, point to things, encourage (even reinforce imitation), and teach them special skills—all of which involve a referential triangle between human, ape, and some third entity."[33] When in close contact with humans, it appears that chimpanzees can gain some access into our social world and even our symbolic world. Training apes to learn some aspects of American Sign Language illustrates this point. Their achievements in these settings, however, do not alter the conclusion that cultural learning is an emergent and distinctive feature of human evolution. Only human beings invent elaborate cultures that their offspring learn through immersion as part of the natural course of cognitive development.

Episodic Culture

How, then, have the capacities of invention and social learning changed across phylogeny in evolutionary time? How have such changes affected the kinds of cultures that we observe in the historical record? Archeologists have studied cultural changes in terms of the artifacts created by early hominids. Change in the sophistication of stone tools is a good example. However, it is also instructive to speculate about the cognitive capacities that have evolved in modern humans. Both cognition and culture set us apart from all other species in biological evolution and, at the same time, characterize the human journey through historical time, from the Upper Paleolithic to today.

The cultural world of the chimpanzee provides a clouded glimpse, perhaps, into the mind of our common ancestor from five to seven million years ago. The neocortex of the chimpanzee is well developed and no doubt serves the function of conscious awareness, just as it does in human beings. However, unlike human beings, the ape's mind, and the mammalian mind more generally, would be limited to representations of its perceptions and actions. Lacking language, an ape cannot name and reflect upon abstractions about what it sees, hears, and feels. The thought of apes is tied to the concrete situation, the here and now of perception and action, rather than the abstract symbolic representations. The ape also cannot talk to itself, using language as a means of inner dialogue and contemplation. The behavioral traditions that apes invent and learn are linked to the episodes that they experience from day to day. Because grooming and backscratching are common episodes in a chimpanzee's life, it is not surprising that the methods used become traditional. The tools invented to fish for termites show the impressive problem-solving capacities of the chimpanzee, but, again, they arise from concrete experiences and reflect what is called episodic culture.[34]

If we assume that the common ancestor of chimpanzees and humans exhibited episodic culture, then what changed in hominid ancestors that led to the culture of modern human beings? It has already been argued that an advanced form of working memory and advanced social intelligence paved the way in part. However, an even more radical change also occurred—the invention and propagation of language within a human population and from one generation to the next. When human beings were bestowed the gift of naming, we opened a door to a world of abstraction that is closed to our primate ancestors. To name meant the capacity to utter not just a sound but also an abstract symbol that referred to something entirely and arbitrarily different from the sound itself. The ability to learn and use a name to refer to a concrete object by itself moved human beings beyond the episodic world of perception and action into a world of abstraction.

There is good reason to see the highly developed social intelligence discussed in this chapter as the foundation for the development of language. Consider an experiment in which a human points to or gazes in the direction of hidden food in the presence of a chimpanzee. Chimpanzees, as well as all other nonhuman primates, "show little spontaneous skill at using such communicative cues to find the hidden food in this cooperative context."[35] Nor do they succeed when a member of their own species provides the communicative cues to help locate the food reward. If trained successfully after dozens of trials to use one such cue, the chimpanzees fail to use even a very similar, but slightly modified, cue of communication. It appears that the social intelligence of nonhuman primates does not enable them to learn how to communicate to solve problems, like finding food.

Armed with the innovative abilities of human working memory and the social learning and cooperation abilities of advanced social intelligence, human beings acquired the third part of the modern ensemble. Language allowed human beings to share their mental experiences with others around them who spoke the same language. Telling stories about episodes became possible in addition to actually witnessing an episode. Language also allowed the possibility of self-narration, or talking to one's self. An ongoing, incessant commentary on current perceptions or episodes remembered from the past or imagined in the future became a defining feature of the human mind. The invention of language opened the door to radically different kinds of cognitive cultures that characterized human beings for millennia up to the present day.

Chapter 4

LANGUAGE

Of all the parts of the ensemble hypothesized here, language *almost* singularly creates an unbridgeable chasm between the mind of *Homo sapiens* and that of *Pan troglodytes*. As noted in chapter 1, Charles Darwin made this point over 150 years ago when he observed that our higher cognitive faculties are "mainly the result of the continued use of a perfect language."[1] Yet the hedge "almost" and the interaction of language with other ensemble parts should not be overlooked in understanding why the human mind is of a fundamentally different kind from anything found in nonhuman primates.

Language provides human beings with an ability to communicate with one another, but it is far more than a system of communication. Language is also a means of representing reality. Through language the mind found a means to represent objects and events in a symbolic manner rather than through visual-spatial images. Because of language, the ancestral mind was no longer tethered to thinking about concrete objects and events that could be imaged like a picture—a reuse of vision for seeing things in the head rather than in the immediate environment. The capacity for symbolic thought—the use of words to refer to objects, events, thoughts, and concepts—is the essence of our species. Not surprisingly, then, the beginnings of language were portrayed as part of the beginnings of humanity in the creation account of the Bible: "And out of the ground the LORD God formed every beast of the field and every fowl of the air; and brought them unto Adam to see what he would call them: and whatever Adam called every living creature, that was the name thereof" (Genesis 2:19).

THE MOTHER TONGUE

The hypothesis that modern human beings evolved first in Africa and then spread through migration to the Middle East and later throughout the world rests on both genetic and archeological evidence. Linguists have long speculated that a protolanguage once existed from which all languages subsequently evolved through historical time. Languages change relatively rapidly over time through the fast-track process of cultural evolution. Across hundreds, if not thousands, of generations, language systems appear to have grown more and more complex and differentiated in a manner analogous to the speciation of plants and animals. As with biological species, languages can also become extinct, dying off with the subpopulation of human beings who know the language. When a language is no longer taught to the next generation, it passes away.

That all human beings once spoke the same language was again an echo of the biblical book of Genesis:

> Now the whole earth had one language and one speech. And it came to pass, as they journeyed from the east, that they found a plain in the land of Shinar, and they dwelt there. Then they said to one another, "Come, let us make bricks and bake them thoroughly." They had brick for stone, and they had asphalt for mortar. And they said, "Come, let us build ourselves a city, and a tower whose top is in the heavens; let us make a name for ourselves, lest we be scattered abroad over the face of the whole earth." But the LORD came down to see the city and the tower which the sons of men had built. And the LORD said, "Indeed the people are one and they all have one language, and this is what they begin to do; now nothing they propose to do will be withheld from them. Come, let Us go down and there confuse their language, that they may not understand one another's speech." So the LORD scattered them abroad from there over the face of all the earth, and they ceased building the city. Therefore its name is called Babel, because there the LORD confused the language of all the earth; and from there the LORD scattered them abroad over the face of all the earth. (Genesis 11: 1–9)

The tower referenced in Genesis was in all likelihood a staged temple tower, known as a "ziggurat," and the "Tower of Babel" could only be the ziggurat erected in Babylon.[2] Babylonian literature refers to the construction of a large ziggurat known as Etemenanki, which was described as a temple to the god Marduk that would reach to the heavens. Archeologists have discovered a tablet giving the precise dimensions of each of the structure's seven stories, and the Greek historian Herodotus provided a description of Etemenanki after his visit to Babylon in 460 BCE.[3]

Theologians and scholars of language have for centuries sought to discover the mother tongue of all humanity, the source from which existing languages descended through cultural evolution in historical time. However, interest in the topic exploded in scientific circles after the publication of Darwin's *Origin of Species* in 1859. Questions about the origin of human nature—including languages—proved irresistible. Yet there was little if any tangible scientific evidence that could be advanced to test theorists' speculations. And so, by 1866, the *Société de Linguistique de Paris* was obliged to ban all discussions of the evolution of language from their meetings.[4] This indeed chilled debate, for it was not until more than a century later that cognitive scientists returned in earnest to investigate the topic.

One intriguing hypothesis is that a common prehistoric language was spoken in Asia around fifteen thousand years ago by modern humans living in the territory between the Black Sea and the Caspian Sea.[5] Called Nostratic— after the Latin *noster* for "our"—it was hypothesized to be a superfamily of languages, including the family of Indo-European languages as well as other linguistic families such as the Afro-Asiatic languages spoken by, for example, Ethiopian and southwestern Asian populations and the Altaic languages spoken by Korean and Japanese populations. Linguists sought to reconstruct the proto-languages or a hypothetical ancestral language from which modern language families descend (e.g., proto-Indo-European). The reconstruction of a proto-language involves specifying the set of consonants and vowels it used, the words it employed, the grammatical endings appended to nouns or verbs, and even the allowable word orderings, such as whether an adjective precedes or follows a noun.

Instead of trying to reconstruct proto-languages, the linguist Joseph Greenberg attempted to build a taxonomy of existing languages by comparing their similarities and differences. He selected words that change the least over time, such as the numbers one, two, and three, parts of the body, personal pronouns, and so on.[6] The picture that emerged from Greenberg's classification turned out to be in many ways similar to the Nostratic hypothesis. Greenberg concluded that a Eurasiatic superfamily included the Indo-European and Altaic languages—essentially all of Europe and Asia along with Eskimo-Aleut in the American Artic. It differed from Nostratic by excluding the Afro-Asiatic family, which Greenberg concluded had split from and evolved as a separate lineage at an earlier point in time.[7]

Greenberg's classification of languages shows that Africa contains within it the greatest diversity of languages. There are four highly diverse families of languages in Africa alone, whereas the entire European peninsula is coextensive with all of Asia in spawning a single family of languages—Eurasiatic.[8] In fact, this same family extends around the world, across the Alaska and the northern zones of Canada, and as far east as Greenland. In much the same way, a single family of languages known as Amerind covers the vast majority of both North and South America. In short, all the languages of Europe, Asia, and the Americas taken together can be classified within only two families; by contrast, the continent of Africa alone has twice that many families. This pattern of greater linguistic diversity within Africa suggests that languages have been evolving for a longer period of time there than anywhere else. Just as greater genetic diversity implies an African origin of modern humans, the diversity of human languages suggests the same conclusion.

Luigi Luca Cavalli-Sforza and his colleagues explored the parallel between genetic and linguistic evolution in fine detail. Might the picture of human origins and migration out of Africa to the Near East and beyond, indicated by similarities in the genomes of diverse human populations, also be supported by comparisons of their languages? The approximately five thousand languages still spoken today can be clustered into seventeen different phyla.[9] The languages sharing a phylum use similar sounding words to refer to the same concept. They are thus related linguistically in a manner that parallels the biological relationships of brothers, sisters, and cousins. For example, Italian, Spanish, and French are related linguistically because they descended from and diversified around the common ancestor, ancient Latin. The DNA results have shown a major split in the phylogenetic tree between Africans and non-Africans. The second split separates, in terms of their genetic similarity, the Caucasoids, East Asians, Artic populations, and American natives, on the one hand, from the Southeast Asians, Pacific Islanders, New Guineans, and Australians on the other.

Strikingly, the linguistic similarities confirm the DNA evidence, almost without exception. Each and every linguistic phylum maps onto only one of the six major clusters of human beings classified on the basis of genetic similarity.[10] For example, Niger-Kordofan, Nilo-Saharan, and Khoisan linguistic phyla are all linked to biologically defined populations in Africa. The Indo-European languages, out of which English descended, are tied to the Caucasoid populations, as defined by their genetic similarity. The few exceptions that failed are not too difficult to explain from what is known about the biological history of the groups in question. For example, Ethiopians speak Afro-Asiatic languages found in North Africa and the Middle East, where populations are genetically identified as Caucasoid. But genetic mixing of the African and Caucasoid lines is also evident in the Ethiopian population. Similarly, the Tibetans are closest to the Northeast Asian cluster in terms of genetic similarity, but linguistically they align with the Sino-Tibetan phylum spoken in all of China. The Tibetans are thought by historians to have originated from nomadic groups from the steppes north of China, a fact that could account for their connection to the Northeast Asian genetic cluster.

That languages are evolutionarily related to one another was suggested first by Sir William Jones in the eighteenth century. As an avocation, Jones mastered and classified on the basis of their similarities numerous languages while serving as a Supreme Court justice in India. In 1786, in his *Third Discourse to the Asiatic Society*, he proposed that Greek, Latin, and Sanskrit, the language of Northern India from the first millennium BCE, were remarkably similar and must have descended by modification from a common ancestor.[11] This linguistic insight came seventy-two years before the explanation of biological evolution in the *Origin of Species*, but it was Darwin's book that was widely read and spread throughout the culture.

DEFINING LANGUAGE

At the core of language is the human capacity for symbolic reference: the use of an arbitrary sound to refer to something else. A word, a sound or combination of sounds, can refer to a concrete object that exists in the external world, such as a dog. Although a seemingly simple example of symbolic reference, it reveals a profound capacity for abstract thought. It reveals an ability to think with symbols rather than with visual-spatial images that are grounded in the objects and events of the physical world. We can think visually about objects and events that have been perceived in the real world. This capacity is certainly shared by other primates endowed with a visual system similar to our own. Yet human beings alone can also think using words that refer to objects and events. *Dog* can be a constituent of thought without any imagery of a particular dog. In short, human beings can think symbolically and abstractly, as seen in the example of naming.

As a symbol, a word is purely an abstract carrier of information, an arbitrary but agreed upon convention among the speakers of a language, a tool to help one think of a particular concept when the word is heard and comprehended. Where speakers of English say "the dog," speakers of German (*das Hund*), French (*le chien*), Spanish (*el perro*), or Italian (*il cane*) use different combinations of sounds to mean the same thing. The arbitrariness of the connection between symbol and referent is all the more striking because English is closely related to both German and French. Similarly, French, Spanish, and Italian are closely related Romance languages with their origin in Latin.

The concept itself is also an abstraction in that a wide variety of four-legged mammals that bark, sleep, wag their tails, and live with human beings all fit the word dog. The reference or extension of the concept in the external world includes a very large and diverse population of concrete entities that we can be see, hear, smell, pet, and at times be bitten by. But the meaning of the concept—what philosophers call its intension—is an abstraction that the mind represents and indexes to all dogs and only to dogs. Such conceptual capacity provides a second example of the human ability for abstract thought.

A final example is that some words refer to concepts that do not have a well-defined extension that can be enumerated. Take, for example, the concept of liberty. The concept itself seems to exist chiefly in the minds of human beings. Its referential extension is as abstract as the concept itself. One can imagine counting up all the dogs in the world today—an onerous task to be sure, which would yield a very large number, but at least it is a finite number. By contrast, it is difficult to conceive of enumerating all possible examples of liberty. A richly diverse and seemingly infinite set of exemplars all fit the word well. Or consider imaginary objects. Lilliputians, hobbits, unicorns, fairies, and the like are all perfectly useful concepts with an indefinitely large extension. But they can be seen only with the mind's eye. They cannot be tracked down and counted in the physical world.

Thus, language uses abstract symbols to convey meaning. Humans use words, or patterns of sound, to refer to objects, events, beliefs, desires, feelings, and intentions. The words carry meanings. If your friend says he is happy, then you interpret this to mean something about his emotional state. If, instead of speaking, your friend whistles a tune, then his behavior may say something about his emotional state, but it is less meaningful. Your friend might whistle by habit or whistle when he is angry, sad, or happy. Unlike speech, whistling is not specialized to convey a clear meaning. Once humans learn a word, they can retrieve its mental representation, hold it in working memory, and use it in thought. The word itself is represented separately from the object or event to which it refers.

At the core of language, then, is an ability to use symbolic representation both in communicating with other human beings and in thought. As Terrence Deacon put it in *The Symbolic Species*, "language is not merely a mode of communication, it is also the outward expression of an unusual mode of thought—symbolic representation . . . symbolic thought does not come innately built in, but develops by internalizing the symbolic process that underlies language."[12]

The symbols of language are invented by human beings in such a form that they can be passed down to the next generation. Language thus must be learnable by children. Although languages can be remarkably complicated, they cannot be so complicated that children fail to acquire them and pass them on to their children. As with other cultural artifacts, language can be modified so that is evolves over multiple generations. For example, after a bit more than four hundred years, or twenty generations, the Elizabethan English of Shakespeare's plays is recognizable but distinct from the British English spoken today. Modern readers of Chaucer's poetry in Middle English, from the fourteenth century, must struggle still more to recognize their language, while the Old English poem *Beowulf*, from the early eighth century, strikes the eye and ear as a foreign language.

The speed of linguistic change through historical time can be dauntingly fast. Merritt Ruhlen illustrated this point in *The Origin of Language* with the example of words used to express approval in American English with "the succession of 'neat' (1950s), 'cool' (1960s), 'bad' (1970s), 'rad' (1980s), and 'awesome' (1990s)."[13] All of these were existing words in the language ("rad" being an abbreviation of "radical"), but entirely new words can also be invented and find their way into the language community. For example, our modern word *dog* came from a new word, *docga*, invented in the Old English of one thousand years ago. This novel entry into the language competed with a much more ancient term found in Proto-Indo-European, whose modern equivalent is *hound*. Over the past millennium, *dog* eventually replaced *hound* as the everyday label for this commonplace category of animals. Although the *hound* remained in the language, it evolved over the generations to a narrow meaning of "hunting dog."[14]

Language can be used to communicate factual information, but this is not its sole function, and it may not even be the motivation for its origin. Venting emotions, joke telling, and social greetings all are common uses of language that bond people together in a world of shared experience, yet they are not aimed at the articulation of facts or even opinions. Moreover, some factual knowledge cannot easily be rendered in language. For example, try to describe to a friend how to tie a necktie without resorting to gestures or drawing a diagram. Or try instructing someone, in words alone, how to navigate by walking or by car to a location in a large city a mile away. Now how much easier is it if you can point to a map? These are examples of visual-spatial knowledge that is not readily encoded in words alone.

Emotions, too, can be too dense to find clear expression in strings of words. All human beings can quickly communicate joy, sadness, fear, or surprise through inborn species-specific facial expressions. These facial gestures are universally understood across all human cultures. Yet sharing these emotions clearly through language challenges the average speaker; it is more a task for the advanced skills of poets. Consider romantic love. How does one say "I love you" in a way that it is understood through language alone, without all the other behavioral ways that human beings express their love for one another? Countless love letters, sonnets, and songs show that even the most skilled with language struggle with the expression of love.

By combining the symbols of language, any human being can say any number of things—the scope of possible expressions is inexhaustible. Consider, for example, a six-word sentence. Suppose that one were to select one of ten possible words for the first word of the sentence, one of another set of ten possible words for the second word, and so on. The number of unique sentences that could be generated following this procedure would equal 10^6 or one million sentences. Because you are not limited to only six-word sentences or to ten possible choices, the number of unique sentences that you might utter is infinite. Thus, abstractions of language are profoundly productive—through them the human mind can generate novel sentences endlessly.

THE MACHINERY OF LANGUAGE

How then is the productive capacity of language realized? The first part of the machinery of language is semantics, or the meanings of all its symbols. A whole word is constituted from brief fragments of sound that distinguish one meaningful word from another. For example, *bill* and *kill* convey different meanings because they differ in their initial sound. Similarly, *braised* and *praised* also differ, but here the difference in pronouncing *b* versus *p* is a very subtle one. The sounds are so close that languages related in their historical evolution often adopt only one or the other to mark meaning shifts. To illustrate this comparison, "language A might have *aba* for 'fish,' *uda* 'walk,' and *paga* 'tree,' while the closely related languages B, C, and D all have *apa*, *uta*, and *paka* for the same three words."[15]

These sounds are the building blocks of meaningful components, the morphemes. A *morpheme* is a minimal unit of speech used repeatedly in a language to code a specific meaning.[16] A word such as *bill* is a morpheme, but so, too, are prefixes and suffixes, such as *pre-* and *-es*. Each morpheme signals a distinct meaning. The suffix *-ed* on the end of a verb tells us that the action took place in the past. So, the word *billed* is composed of two morphemes, each of which conveys a specific meaning.

All the morphemes in a language, taken together, make up a mental lexicon, or the dictionary of long-term memory that humans rely on in speaking and listening and in reading and writing. Each morpheme is a lexical entry in this dictionary of the mind. In particular, one is concerned in semantics with content words, that is, the verbs and the nouns. Function words, such as articles (e.g., *the*) and prepositions (e.g., *by*), often serve more of a grammatical function rather than a semantic role. For example, the statement "The telephone company bills by the month" would mean the same thing if the prepositional phrase *by the month* were replaced by the adverb *monthly*. However, if *month* were replaced by *week*, then the meaning of the sentence would change to a more frequent billing cycle. Replacing the content word *telephone* with *electric* changes meaning, too, altering where one should send the payment.

The sound of a language is influenced by its phonology, the sound segments that make up the words. Some phonological segments make a difference in meaning in a given language, and these are called *phonemes*. English makes use of forty-six phonemes, some of which are used in other languages and some of which are not.[17] These include the consonants, the vowels, and combinations of vowel sounds known as *diphthongs*. For example, the consonants *l* and *r* signal different morphemes or word meanings, such as the difference between *look* and *rook* or between *lip* and *rip*. However, in Japanese this phonemic distinction is not made and the two sounds can be easily confused when native Japanese speakers learn English. Similarly, there are sounds in other languages that are not employed in English. In Spanish, the rolled *r* is articulated near the front of the mouth and is slightly different from the rolled *r* of French, which is articulated further back. Neither of these methods of articulation is used in pronouncing an English *r*.

Languages also differ in the sequence of phonemes that are permitted. In English, one never sees the sequence *pt* at the beginning of a word, whereas in Greek the combination is common, as can be seen in borrowed words, such as *pteropod* or *pterosaur*. The rules of phonology are learned implicitly through repeated exposure to a particular language. Although we are not consciously aware of these rules, we can easily decide whether a nonsense word is possible (*patik*) or impossible (*ptkia*) by making use of what we have learned unconsciously about English phonology.

The sound structure of a word is detached from its semantics in terms of the location in the brain where each is processed. In the inferior region of the left prefrontal cortex, one module specializes in processing the meaning of words whereas another is dedicated to their sounds.[18] If you think about a word that is semantically related to another word (e.g., *chair* is related to *table*), or think about whether a word is abstract in meaning (e.g., *freedom*) versus something concrete and easy to visualize in the mind's eye (e.g., *tree*), then a semantic network in the inferior left prefrontal cortex is activated. If, on the other hand, you think about how many syllables a word has or whether it rhymes with another word, then a distinct phonological network is activated. This phonological network is the same as that found in the phonological loop of verbal working memory. It is located somewhat posterior in the left prefrontal cortex from the semantic network.

Sometimes during speech production, the sound of the word we want to say cannot be retrieved. There is a feeling of knowing what needs to be said, but sound structure is temporarily lost. Bits and pieces might be accessible—such as knowing the first letter or the number of syllables—but the full phonology cannot be retrieved. This frustrating experience is called a *tip-of-the-tongue state*.[19] It illustrates that the semantic representation of a word—its meaning—is stored separately from its phonological representation—how it sounds when pronounced. Unless both meaning and sound structure are jointly retrieved, the word cannot be named.

Spoken language unfolds in a sequence over time. One word follows another in temporal order. Whereas a visual event can be simultaneously grasped as a whole, the sound structure of language or any other auditory source of information, such as music, has a temporal structure. Syntax refers to the grammatical rules that specify how words and other morphemes are arranged in a sequence to yield acceptable sentences.

For example, consider this arrangement of words: "Young children learn language very rapidly." This string of words follows the syntactic rules of English; the sentence begins with an animate agent ("children") that serves as its subject, followed by a verb that describes what the subject does ("learn"), which in turn is followed by the direct object of the verb ("language"). The subject is modified appropriately by placing an adjective ("young") before the subject; reversing this word order creates a syntactic error that sounds peculiar to the ear of the listener ("Children young . . ."). However, there is more flexibility regarding where to position the adverbs; thus, "Young children very rapidly learn language" is also grammatical. The morpheme -s must be added to the verb if the subject is changed to singular rather than plural ("A young child learns . . .") for the sentence to be grammatical.

When children learn their native language, they acquire implicitly an understanding of the grammatical rules and exceptions to those rules. Implicit learning occurs without conscious awareness of the rules. Just as the allowable sequences of phonemes are learned and used unconsciously, so are the permissible orders of morphemes. This refers not just to the order of whole words, but also to the use of morphemes as word endings, such as adding -ed to a verb to express that the action took place in the past. Knowing the syntax of a language implies the ability to decide whether a string of words is grammatical and whether two different sentences mean the same thing. For example, "Language is learned by children very rapidly" is easily recognized as semantically equivalent to our earlier sentence despite the difference in syntax.

Exactly why children are able to acquire implicitly the syntax of their native language with such seeming ease is a controversial issue in cognitive science.[20] The nativist view is that the human genome has encoded neural systems that are dedicated to the acquisition of language. On this view, a universal grammar, providing expectations about the way all languages are structured, and a language acquisition device are innate components of the human brain. Other theorists question the need for these entities and argue instead that general learning capabilities are sufficient to acquire the syntax of one's native language. Despite decades of research by linguists and psychologists, the issue remains unresolved.

The machinery of language is bolted together by the social interactions of human beings. What we say and how we say it is ultimately shaped by those who listen to, comprehend, interpret, and respond to our utterances. For example, if it is too hot in a room you might say to others sitting near the window: "Open the window!" But barking out a command is rude among social equals. So you might instead ask a polite question: "Could you please open the window?" The listeners would not take this as a literal question about their physical or cognitive capacities. Rather, they would infer that you would like the window opened. You might even make a simple declaration: "It sure is hot in here." Although the inference required is a bit more complex, listeners would likely understand that you are really doing more than expressing how you feel at the moment.

Pragmatics is a term used to describe how language use is shaped by its social context.[21] Although we sometimes talk aloud to ourselves or use covert language to think in silence, spoken language generally takes place in a conversation with others who understand and can produce the language. Language is in large measure employed in dialogue rather than monologue. Speakers and listeners take turns and participate in a ritual of give and take, sharing ideas back and forth. Conversing together follows a rhythm of social interaction that bears much resemblance to other joint activities, such as dancing or playing tennis.

Violations of the pragmatic rules of language are as glaringly apparent to us as hearing an ungrammatical string of words that does not fit the rules of syntax. For example, if one speaker in a small group hogs the conversation and refuses to allow others a chance to speak, the failure to take turns is obvious and irritating. Participants agree to say things that are appropriate to the conversation and to end the conversation at a mutually agreeable point. For example, a partner who appears incapable of ending a conversation or who brings up touchy subjects at a party becomes someone to avoid. Speaking audibly in a shared language is required. Imagine a small group of conversationalists in which two of the participants keep whispering to each other or speak in a language only the two of them can comprehend—the couple's secrecy is not amusing to the others. One of the trickiest rules of cooperation is to be informative and truthful. An evasive speaker easily annoys listeners and is not always successful in concealing the truth. But at the same time insulting others with the truth is not socially adept either. Compliments, even when they are little white lies, can be socially useful.

The pragmatic dimension shows us that language is of one piece with the social cognition of joint activities of all kinds.[22] Some joint activities are highly verbal in nature, such as dialog, whereas others are less so, such as playing cards or making love. In all joint activities, the two or more individuals meld their attention into a shared mental space. Although dialog is more verbal than other joint activities, the fact that people share common ground as the basis for their conversation means that utterances can often be remarkably telegraphic in nature. One of the rules of conversation is to say as much as is necessary but no more—there is no need to state things explicitly when they are part of the common ground. For example, imagine a boy and a girl watching a television program together. The boy, holding the remote control, asks: "Boring?" The girl responds: "Switch." Immediately, the boy starts to channel surf. Because the current program was the focus of joint attention, there was no need for the boy to say, "Do you think this program is boring?" And the girl did not need to even respond to his single word question by saying, "Yes, I think it is boring, so please change the channel." In the joint activity of dialog, meanings can be grasped with a minimum of words and no need for even simple syntax. Indeed, had the girl simply rolled her eyes, the boy would probably have understood and switched channels.

THE ROOTS OF LANGUAGE

Humans are not alone in thinking and communicating, as Donald R. Griffin documented throughout his 1984 book titled *Animal Thinking*. For example, a honeybee can direct other members of the hive to distant sources of nectar by waggling its body as a signal. The precise nature of the bee's waggle dance communicates the direction and distance from the hive of a source discovered by the dancing bee. Other kinds of animal communication systems include the antennae and head gestures of weaver ants, the complex sonar signaling of dolphins and whales, and the alarm calls of monkeys. By studying the communication systems of primates, it may be possible to identify the roots of language.

For example, vervet monkeys from Eastern Africa make five different kinds of calls—each warns others in their group to the impending danger posed by one of five different predators.[23] Similarly, Diana monkeys, a type of West African monkey, make distinct calls to signal their distress about different predators. For example, one call is used to warn others of an eagle circling overhead, whereas another different sound is produced when the leopard is approaching on the ground. Yet it is unclear what exactly the monkeys are experiencing when they produce and perceive these alarm signals. Are they just responding to the sounds of the calls or do these calls trigger a mental representation of the eagle versus a leopard? Are they really referring to a specific referent or do they simply signal the importance of orienting to the environment to detect a threat of some unspecified kind? Just because a different signal is used for different predators does not necessarily mean that the producer is mentally representing a specific class of predator or that the listener does so on hearing it. In other words, do these calls truly point to a specific referent?

In a field experiment, researchers played a recording of the sound of an eagle call to wild Diana monkeys. As they would with a real eagle, the group immediately began emitting a high rate of Diana eagle alarm calls designed to warn of the danger overhead. Within five minutes the rate of calling decreased to zero, because they had already warned each other of the presence of the eagle. In fact, if another eagle call is then immediately played, the monkeys do not sound the alarm at all. So, what would happen if a Diana eagle alarm were played first, instead of the call of an actual eagle? Would the monkey understand the meaning of this and promptly join in to warn others? Would they after five minutes then completely ignore a real eagle call? The answer is yes to both questions.[24] The results seem to imply that the Diana eagle alarm call actually refers to a real eagle. In other words, the primate brain shows the rudimentary capacity to use a sign with a specific meaning.

It would be going too far, however, to equate a monkey's predator call with a word. The call is used when the monkey perceives an actual eagle or leopard in the environment. It is used to signal a threat about the here and now. The call is not typically used when the threats are not physically present. It could be that the call is acting as a conditioned stimulus to an unconditioned hide or flee response. By pairing the call with the specific predator present, the monkeys would learn to hide or flee in response to the call alone. If this is so, then it may be that "monkey alarm calls do not refer symbolically to snakes, eagles or leopards, but rather elicit differentially conditioned flee responses associated with the presences of these predators."[25] Words, by contrast, allow symbolic thinking about the concept of an eagle or a leopard anytime, anywhere. The word can be disconnected from an immediate physical referent and employed in thinking about the concept in general. The word is acting as a symbol detached from the referent object in the real world; that is, it refers to a concept in the mind. It is this symbolic power of words that allows us to invent words for abstract concepts that are purely imaginary (e.g., a unicorn or the square root of -1).

There is an interesting exception in that primates employ "deliberate uses of alarm calls in the absence of any predator, designed to distract other monkeys from aggressive intentions or to remove potential competitors for some item of food."[26] The monkeys appear to have learned that an alarm call can be used to deceive others precisely because the species as a whole interprets it as a signal of imminent physical danger. Again, they could have learned how to deceive using the call as a conditioned stimulus. By contrast, we can listen to a story about a leopard without panicking, fearing an imminent attack, and fleeing. The human word *leopard* is clearly acting as a symbol for a concept.

What, then, can be said about the language capabilities of chimpanzees? Does the 2 percent or so difference in our genomes preclude language in non-humans? The focus of research has not been on the communication abilities of wild chimpanzees, but rather on attempts to teach chimpanzees language in laboratory environments. This line of research has a long history, with early efforts dating to the 1930s. Winthrop and Luella Kellogg attempted to teach a chimpanzee to speak by raising it with their own son and treating the chimpanzee as a human being. The chimpanzee, named Gua, was not explicitly trained, but rather was expected to learn by observation right along with their son, Donald. As Roger Fouts recounted the history in his book *Next of Kin*: "Unfortunately for science the Kelloggs abruptly terminated the study because, rumor has it, Mrs. Kellogg became distressed when Donald began acquiring more chimpanzee sounds than Gua was acquiring animal sounds."[27]

One reason for the Kelloggs' failed effort likely was the superior skill at imitation shown by human beings compared with chimpanzees. However, an inherent limitation in the vocal track of the chimpanzee to produce the phonemes of human speech was fatal to the project. Unlike parrots, which can mimic human speech sounds to a degree, apes cannot match the full range and felicity of human speech production.

The difficulty lies in the location of the chimpanzee's larynx.[28] The human larynx, where the vocal cords are located, begins at birth in a high position in the vocal tract, comparable to the position found in the chimpanzee. However, beginning at about three months of age in a human being, the larynx migrates lower in the neck into the pharynx, assuming its adult position around fifteen years of age. For all mammals except human beings, the laryngeal tract that connects the nasal passages and mouth to the lungs is clearly separate from the pharyngeal tract that connects the mouth to the esophagus on its way to the stomach. Mammals can eat and breathe at the same time without worries of chocking on food or drink. In the design of human beings, this obvious advantage for the survival of the species was given up in order to position the larynx low in the throat. Why? Because the precarious position of our larynx permits the articulation of an astonishing range of sounds that constitute the phonemes of language. Without it we could not articulate the three vowels *i*, *u*, and *a* or the consonants *k* and *g*. The larynx starts at birth in a high position that permits a new born to nurse and breathe simultaneously without danger of choking to death. The larynx then descends, and by the age of three months the infant cannot eat and breathe at the same time. But the infants capacity to babble all of the possible phonemes of all possible human languages begins to emerges along with its descent. Speech is such a powerful adaptation that *Homo sapiens* risk asphyxiation to attain it.

The low position of the larynx is not the only biological adaptation for speech. The chimpanzee lacks the extraordinary vocal control of human beings. In other words, the brain's motor control of the lips, tongue, jaw, and other components of sound articulation differ markedly in the two species. A critical gene for speech motor control, called FOXP2, has in fact been identified. By studying the genetic basis of a disorder of the face and mouth muscles, scientists were able to identify a mutation of the FOXP2 underlying the disorder.[29] In a particular family, across three generations, half of the members suffered from the disorder that severely impaired their speech because the oral movements required for articulating the phonemes of speech could not be controlled. The disease starts early in childhood and impairs not only the normal progression of learning and using oral language but also the language and grammar skills that come later, including writing. The inheritance pattern revealed a single mutated gene that disrupted the development of the basal ganglia in the telencephalon of the brain that are known to be crucial for motor control.

The FOXP2 gene is also found in other primates. However, it is a slightly different variant in humans compared with nonhuman primates. Without this variant intact in human beings, there is severe impairment of the motor output of speech. Thus, lacking the right larynx and control of sound articulation, the chimpanzee physically cannot imitate human speech—the precise motor control is lacking. This difference between chimpanzees and human beings suggests that "some human-specific feature of FOXP2 . . . affect a person's ability to control orofacial movements and thus to develop proficient spoken language."[30]

Given these facts, scientists then turned to nonverbal forms of language and achieved considerable success relative to the early work by the Kelloggs. One project taught a chimpanzee named Washoe American Sign Language (ASL). By training Washoe to make the appropriate sign when shown a picture of an object, she was able to learn to express 132 different ASL signs and comprehend many more than that.[31] The expressive vocabulary learned by Washoe seemed to match reasonably closely with the words uttered by young human children. For example, Washoe learned to use a large number of category names, such as flower, fruit, and cat. In a different project, a bonobo named Kanzi learned how to choose visual symbols on a computer display called lexigrams. Each lexigram referred to a specific object or action. Kanzi learned from watching his caretakers point to each symbol appropriate to an object while speaking to him about "daily routines, events, and about comings and goings at the laboratory" and also about "trips to the woods to search for food, games of tickle and chase, trips to visit other primates at the laboratory, play with favorite toys such as balloons and balls, visits from friends, watching preferred TV shows, taking baths, helping pick up things, and other numerous simple activities characteristic of daily life."[32] By the age of six, Kanzi could identify 150 lexigram symbols on the computer when he heard the words spoken to him; he could also perform correctly 70–80 percent of the time in comprehending and responding to novel sentences, such as "Put the rubber band on your ball" or "Bite the stick."[33] By way of comparison, a mother taught her human infant the lexigram board, along with the usual spoken language. It turned out that Kanzi did just as well, if not slightly better, than the child was doing at two years of age.

The findings of ape language learning were greeted with considerable skepticism, and they generated deep controversy. Arguments ensued regarding whether the hand signs or lexigrams were truly operating as symbols in the apes as they do in humans. Others contested whether the apes had knowledge of syntax in their comprehension of novel sentences or in their production of short sequences of signs. Certainly, it was apparent that their maximum vocabulary size was small and that their ability to produce novel expressions was highly limited. A hallmark of human language is its productivity—the ability to use a large but finite number of words to generate an infinite number of acceptable sentences. Against this standard, the apes fell far short. Their vocabularies are not just small; they also lack grammatical items and show no signs of sophisticated grammar such as embedded clauses.[34]

The divide between chimpanzees and humans in language capability should not be surprising. Chimpanzees have been evolving for approximately six million years since the last common ancestor hypothesized by evolutionary biologists. That there is any evidence at all of symbol use and semantic reference learned by chimpanzees in laboratories is surprising. So is the possibility that the alarm calls of monkeys are indeed primitive forms of semantic reference. Such findings hint at just how ancient the basis of semantics might be. What has become an exquisite system of symbolic thought and language use in modern humans may well have taken a very long time to develop. The adaptations that allow for the gift of language in humans are many and complex. That they were a long time in coming ought to be expected.

What, then, is known about the biological adaptations for language that are part of the human genome? With the exception of FOXP2, the genetic determinants of language remain unknown. But there must be other important genetic mechanisms yet to be discovered. For example, it is known that the brain develops specialized structures for the production and comprehension of language. Most people are right handed, which means that the left hemisphere is considered dominant and contains a zone specialized for language. The language zone incorporates numerous brain structures in the left hemisphere, and damage to any of them can impair language use.

Among the most critical and best known are the following regions.[35] Broca's area is located in the left frontal lobe and plays a critical role in the motor control of speech and in processing the syntax of a sentence. Damage to Broca's area causes difficulties with producing grammatical speech. Such injuries disrupt the ability to understand language when grammatical information is crucial. For example, "playing the field" and "the playing field" are one and the same to the right hemisphere. Without Broca's area, in the left hemisphere, the difference in meaning of these two phrases is lost. For a patient diagnosed with Broca's aphasia, a sentence such as "The boy kissed the girl" is difficult to interpret. Who was kissed? Without an intact Broca's region, the sentence cannot be correctly parsed to know that it was the girl who was kissed. Near Broca's area, only lower and a bit further forward, is a region that specializes in the phonological analysis of speech. Wernicke's area is located in the parietal lobe, near its boundary with the temporal lobe; it specializes in the comprehension of words. Patients with injuries to Wernicke's area have no trouble producing grammatically complex and highly fluent language. Their syntax is fine, but there is a problem in the semantics. Sentences, though fluent, can be vacuous in meaning or difficult to understand because of difficulties with the semantic reference of the words. At times, neologisms or made-up words are added. Finally, the left angular gyrus lies adjacent to Wernicke's area, a bit further back in the parietal lobe, near the occipital lobe. This region processes the visual features of written language, known as orthography, as opposed to the sound, or phonology, of language. Injuries to this area can disrupt reading and writing even though spoken language is unimpaired. The specific roles of genes, and the ways in which they interact with the social environment to produce such specialization of the left hemisphere for language, are still today unknown.

ORAL CULTURE

The monumental fact of history—indeed there would be no history if written language had never been invented—is that human beings alone ventured into a cultural world permeated by language. Our genes prepared us for an entirely new voyage of cultural evolution. The importance of this singularity of nature cannot be overstated. Although human beings have many distinctive features of body and mind, language stands out as the single most significant, for it lifted us into a realm of symbols and abstract thought. The civilizations of human history would never have been possible without the invention of language.

The Upper Paleolithic culture of early modern human beings was characterized by well-crafted stone tools. But it was the art that clearly documented the capacity of early modern humans for symbolic thought. Merlin Donald asked in *Origins of the Modern Mind*:

> What sort of adaptation could possibly explain the explosion of tools, artifacts, and inventions of all sorts for all sorts of applications, and the eventual creation and maintenance of tribal political and social structures, which regulated everything from marriage to ownership, from justice to personal obligation. What change could have broken the constraints on mimetic culture, leading to the fast-moving exchanges of information found in early human culture? Speech and language are the obvious candidates to single out for these roles.[36]

The most important way that language altered the cultural landscape was not in the making of tools and weapons. Instead, it allowed the development of a richer social structure because people had a way to communicate their thoughts directly with others in the tribe. This greatly magnified the ability to transmit cultural knowledge from one generation to the next. Parents and elders could not only show the young, they could tell them. The culture became oral such that vast amounts of what a group knew could be told as narrative. What had happened in the past became known in the present through the power of language. By telling the stories of the past, the past could be preserved and its wisdom passed on to the next generation.

Before the invention of language, did our hominid ancestors communicate in some other way? If that were so, then could there still be some behavior of contemporary human beings that preserves this earlier mode? One intriguing candidate is the use of gestures and poses, the arts of nonverbal communication such as dance and mime. Through facial and other bodily expression human beings are certainly capable of nonverbal communication. Emotional states certainly can be very well communicated in this manner, but other states of mind can be shared among people, at least in a skeletal form, without the use of words. A mimetic form of culture—one based on bodily expression—may have preceded the invention of language and oral culture. It may have been an aspect of the culture of *Homo erectus* that today exists in the behavior of modern human beings as something of a cultural relic or fossil.[37]

Consider all the ways that a mimetic capacity is still with us. Dance is a powerful means to express feelings, and it can, with professional skill, do so in a highly sophisticated manner. A professional mime can communicate a complex sequence of events, accurately and often humorously, with nothing but adept bodily expression. Although we differ in skill and practice levels, each of us can pantomime to some extent when playing a game of charades. Stage and screen actors learn that their capacity to express emotions and thoughts through bodily expression is as important, if not more so, then their vocalizations.

On an everyday basis, human beings use body language to communicate their fears, hopes, and desires with one another. Our bodies speak to others, sometimes conveying the same message as our spoken words and other times belying them. These bodily expressions can be amorphous and not easily read across cultures, but the fact that they exist at all is the significant point here. By contrast, the facial expression of basic emotions is a universal form of human communication. The hand, arm, head, and eye gestures that often accompany human speech are particularly interesting in the way they help listeners to understand. They can convey some information about the speaker's thoughts that is left unspoken. Such gestures are thus part of the integrated system of communication.

Moreover, speech gestures also seem to help the speaker as well as the listener. The gestures represent some of what a speaker intends to say, thus freeing working memory to concentrate on the thoughts that will be expressed through speech. Experiments have shown that when speakers spontaneously gesture while they talk, they are able to perform a second concurrent task better than when they choose not to gesture or are instructed not to gesture.[38] The gestures that accompany speech reduce the load on working memory in much the same way as externalizing thoughts by writing them down. Put simply, spontaneous gesturing helps the speaker to think.

The power of gestures in human communication is seen most powerfully in deaf children. They are able to communicate successfully, even if they have not been exposed to sign language, by inventing iconic gestures that allow others to get the picture intended. These gestures, like those of American Sign Language and other formal nonspoken languages, assume "not only the function of language but also many of its formal features, such as segmentation (producing separate gestures to represent objects and the relations among them), combination (combining those gestures in a structured manner), and recursion (producing more than one proposition within a single gesture sentence)."[39] The expression of thought through language erupts from the modern mind, as urgently through the hands as through the mouth.

Oral culture is the medium by which human beings have made sense of the world since the origin of the modern mind. It has provided the communal forum for the building of conceptual models—such as those of religion and science—and for the composing of stories regarding our identity, history, and destiny. What are we to make, then, of the human need to make sense of the world? Is the human drive for explanation simply a consequence of the invention of language and a product of oral culture? According to the ensemble hypothesis advanced here, there is more than just language at work in this drive. It was instead a capacity to narrate events as inner speech and infer the causes of those events that lies behind the human drive to interpret all that happens to us.

Chapter 5

THE INTERPRETER OF CONSCIOUSNESS

As useful as language is for communicating with others, it is also a means for silent thinking and internal dialogue. Human consciousness is unique from that of any other species because we can talk to ourselves. The mind spins a yarn that explains to us, through the voice of inner speech, our experiences in the world. This inner story telling or narrator is known in the scientific literature as the interpreter of conscious experience, and it has been shown to reside in the left cerebral hemisphere of the brain.

The discovery of the interpreter began with the famous split-brain research of Michael Gazzaniga and his colleagues. They studied epileptic patients whose severe seizures, uncontrolled by medications, were successfully treated by surgically separating the two cerebral hemispheres. Neurosurgeons cut the fibers of the corpus callosum, a massive band of fibers in the interior of the brain that connects the left hemisphere with the right hemisphere, with the aim of stopping the electrical storm of haphazard neural firings underlying the seizure. Not only did the operation work, the patients appeared to show no adverse side effects, at least as detectible in everyday cognitive functioning. In special laboratory tests specifically designed to reveal the left and right hemispheres functioning independently of each other, some remarkable findings emerged.

In a typical experiment, the patient was seated in front of a screen onto which words were projected briefly.[1] Each trial began with the individual being told to fixate his or her gaze on the dot projected in the center of the screen. Next, a word was flashed for only a tenth of second or so either to the left of the dot or to the right. The researchers capitalized on the fact that all stimuli in our left visual field are processed first by the right hemisphere before being sent via the corpus callosum to the left hemisphere. Conversely, in a normal participant, any word presented in the right visual field is first perceived by the left hemisphere. In the case of the split-brain patient, what entered the right hemisphere stayed in the right hemisphere, having no easy means of transit. The same was true for a word perceived only by the left hemisphere because the corpus callosum had been severed.

When the word *ring* was flashed on the right, the left hemisphere was able to recognize the object and to name it promptly using its language capacity. On the other hand, when the word *key* was flashed to the left hemisphere, the patient said nothing. If pressed as to what word had been presented, the patient said "I don't know." The right hemisphere, though, is known to possess the capacity to recognize objects. It is in fact highly skilled at the task of perceiving objects in a holistic, rapid manner. Was it possible, then, that the right hemisphere knew more than it could say?

The researchers tested for the unspoken wisdom of the right brain by asking the patient to reach under the projection screen and pick up, one at time, several objects. These test objects could be felt but not seen. As it happened, the patient was easily able to pick out the correct object, in this case a key, if and only if he or she used the left hand, which is under the control of the sensory-motor cortex in the right hemisphere. In other words, the knowledge that the right hemisphere had the object could be expressed through the hand's privileged access to that knowledge. The speech systems of the left hemisphere were in the dark. Particularly compelling, so were the sensory-motor systems of the right hand, mediated as they are by left hemisphere.

Here, then, was an astonishing discovery. It flies in the face of the fact that our phenomenal experience of consciousness is unitary. Attention binds together different streams of information processing, including those initiated within the left hemisphere and those of the right. With the corpus callosum intact, we are normally unaware of the independent contribution of each hemisphere. Yet the split-brain findings showed that the consciousness of the human left cerebral hemisphere could operate independently of the right hemisphere, at least in patients for whom the normal connections between hemispheres were surgically severed for reasons of medical necessity. The split-brain studies prompted thousands of experiments on normal individuals with intact communication between the left and right hemisphere by measuring brain wave activity and employing other techniques of cognitive science. Brain waves refer to the fluctuations in microvoltage that emanate from the skull as a consequence of neural activity in underlying regions of the brain. They can be detected using an electroencephalograph (EEG) that monitors changes from as many as 128 electrodes positioned all over the skull. The EEG signal picked up from a specific electrode is generated by a population of neurons in an area several millimeters in diameter. The signals generated from a small region of cortex can be accurately measured from millisecond to millisecond through EEG recordings.

Other important discoveries followed regarding the propensity of one hemisphere to be biased in favor of processing particular kinds of analysis.[2] For example, the left hemisphere is biased toward processing language, whereas the right hemisphere is biased toward visual-spatial tasks. The left hemisphere is better inclined to cope with tasks that call for sequential, analytical, logical reasoning, whereas the right hemisphere prefers tasks that benefit from simultaneous, holistic, intuitive judgments. The notion of the left as the rational brain and the right as the creative brain morphed into a pop culture, cartoon version of the scientific facts on hemispheric differences: to become more creative, one should think only with the right brain. The reality is that the brain acts in concert, as a whole, with each hemisphere simply biased toward greater efficiency in one domain over another. Indeed, it is only when the connections between the two hemispheres are severed that one can begin to see the specialized character of right versus left processing. More on point, one cannot turn off the left brain in order to turn on the right brain, nor would one want to do that, giving up fully half of the brain's computational power. To do so is like telling the right half of the orchestra to play while silencing the left.

THE INTERPRETER OF
THE LEFT HEMISPHERE

As famous and influential as the split-brain research became in popular culture, the most significant discovery of them all escaped the notice of the public eye, namely that the left hemisphere serves as an interpreter of the conscious experiences of the brain's cerebral hemispheres. It seeks an explanation of why events occur and concocts a story of the causal relations involved. The left hemisphere uses its linguistic capacity to narrate why such and such occurred. The familiar inner voice that is such an intimate part of the self is the left hemisphere going about its work of commentary and explanation. In Michael Gazzaniga's words, "The interpreter, the last device in the information chain in our brain, reconstructs the brain events and in doing so makes telling errors of perception, memory, and judgment."[3]

To illustrate the interpreter, suppose a command "Take a walk" is flashed in the left visual field to the mute right hemisphere of a male split-brain patient. Although the patient is unaware of seeing the words, he will respond to the command, push back his chair from the table, and get up and walk. Michael Gazzaniga explained:

> You ask "Why are you doing that?" The subject replies, "Oh, I need to get a drink." The left brain doesn't know why it finds the body leaving the room. When asked, it cooks up an explanation.[4]

In a clever experiment designed to catch the interpreter in the act of spinning an explanation, Gazzaniga and his colleagues simultaneously presented a scene of a snowman and a snow covered house to the left visual field and a chicken claw to the right visual field.[5] The patient then pointed with the left hand to one of four test pictures (shovel, lawn mower, rake, pick) that was most appropriate to what had been seen by the right hemisphere (the snow scene). The left hand, controlled by the right hemisphere, was easily able to point to the shovel. In the same manner, when the patient pointed with his right hand to the most appropriate of four choices (toaster, rooster, apple, hammer), it picked out the rooster, what had been registered only in the left hemisphere.

The experimenter then asked the patient why the left hand was pointing to the shovel. Because the right hemisphere is mute, the left hemisphere interpreter had to take charge to respond. Because the interpreter did not know why the right hemisphere picked the shovel with the left hand, it made up a story that fit with its conscious experience of the moment. The interpreter said the right hemisphere selected the shovel to clean out the chicken shed!

In a similar test, a split-brain patient registered the word "bell" in the right hemisphere and "music" in the left. Again, selecting from a set of test pictures, the right hemisphere selected a picture of a bell. When asked why, the interpreter of the left hemisphere concocted a reason that incorporated the information of which it was currently aware: "Music—last time I heard any music was from the bells outside here, banging away," making reference to the bells ringing from Dartmouth library.[6]

The left hemisphere attempts to make sense of the emotional contents of the right hemisphere as well as its cognitive contents. Experimenters presented an emotionally positive stimulus selectively to the right hemisphere, as a means of inducing a positive mood shift within that half of the brain. The left hemisphere begins to interpret its current experience in a positive way, too. What was described moments before as a neutral experience by the interpreter is now reassessed as a positive emotional experience—in this way the sudden shift in the emotional state of the right hemisphere is made sense of by the left's interpreter.[7] In a converse situation, when the right hemisphere is shifted to a negative mood, the left hemisphere now expresses negative feelings instead of what had just been neutral.

Another example of the left hemisphere interpreter can be seen in hypothesis testing. The interpreter seeks explanations for events by formulating and testing hypotheses. Just as a scientist uses hypotheses to explain observations of the world, the left hemisphere does the same to explain its current conscious experience. Consider a test in which a light comes on either at the top or the bottom of a screen on a series of trials. On some trials, a small green screen appears at the top of the screen, while on other trials a small red square appears at the bottom. The participant in the experiment attempts to predict whether on the next trial the light will come on at the top or at the bottom. The prediction is made by pushing a button labeled *top* or one labeled *bottom*.

To test split-brain patients, the lights were flashed to the right visual field and the right hand was used to respond—this assessed the left hemisphere's performance of the task. Or, alternatively, the lights were flashed to left visual field and the left hand was used to respond in order to tap how the right hemisphere performed the task.

In this task, the actual sequence of lights is determined randomly, except that the probabilities of the light appearing at the top versus the bottom are not 50 percent each. Instead, the task is biased so that the top light comes on 60 percent of the time and the bottom light 40 percent of the time. To make the most correct predictions, the optimal strategy would be to guess the most probable response—the top light—every time. That way, it is certain that the prediction would be correct 60 percent of the time—one would maximize performance following such a simple strategy. In split-brain patients, when the task is presented to the right hemisphere, the choices made generally come close to the maximizing strategy.[8] In other words, the right hemisphere is able to perform the task in a manner that comes close to optimal guessing.

The left hemisphere interpreter, however, tries out all sorts of hypotheses about the sequence of lights seen in the past. It assumes there is some underlying explanation for why the lights occur in the order they do and sets about trying to figure out the pattern. As a consequence, the left hemisphere fails to take advantage of the simple strategy of picking the most probable outcome most of the time. Instead, it tries to figure out a rule that explains the pattern underlying the distribution of 60 percent top light and 40 percent bottom light. For example, the interpreter of the left hemisphere might hypothesize that the sequence for five lights is top, top, bottom, top, and bottom. Such a hypothesis is incorrect, since the lights come on at random, with the only constraint that the top light comes on 60 percent of the time. It could have just as easily been bottom, bottom, top, top, and top, which would still fit the 60 percent top rule. Hypothesis testing will inevitably lead to erroneous predictions, because the actual sequence is random. The key point is that the left hemisphere performs worse than the right hemisphere because the interpreter leads to complicated hypotheses that fail.[9] Normal participants in whom the corpus callosum is intact make the same mistake; they, too, come up with elaborate explanations of when the top or bottom light will appear, convinced that they have discovered a pattern in what is random. In the normal human brain, then, the dominant left hemisphere overrules the passive, but in a sense brighter, right hemisphere. Ironically, then, the normal subjects make more mistakes in the task than the left hand of the split-brain subject.

Two strange manifestations of the interpreter come from brain injured patients. Patients with anosagnosia lack knowledge of the fact that they are experiencing a severe neurological problem in the parietal lobe, which keeps track of the position of the body in space, monitoring the location of the left arm, for example. In anosagnosia, the patient denies awareness of the problem. An injury to the right parietal lobe disrupts the monitoring of the opposite side of the body, preventing perception of the left arm's current location. In Michael Gazzaniga's words, "Those who suffer from right parietal lesions and are hemiplegic and blind on the left side frequently deny they have a problem and claim the left half of their body is not theirs!"[10] The anosagnosic patient with a right parietal injury cannot locate his left hand when asked to by a neurologist. The patient says that he does not know where his left hand is now located: "When a neurologist holds this patient's hand up to his face, the patient gives a very reasonable response: 'That's not my hand, pal.'"[11] The patient's explanation is a product of the left-brain interpreter trying to reconcile the perception of the hand with the seemingly deeper belief that the left hand does not exist.

A different kind of neurological disorder—a kind of amnesia—has similarly revealed the interpreter at work. Because of memory loss, the amnesic patient experienced confusion about her current residence. She believed she was in her home in Freeport, Maine, rather than in the Memorial Sloan Kettering Hospital in New York. Michael Gazzaniga recounted her response to his question of where are you as follows:

> "I am in Freeport, Maine . . . I know you don't believe it, but I know I am in my house on Main Street in Freeport, Maine." I asked "Well, if you are in Freeport and in your house, then how come there are elevators outside the door here." The grand lady peered at me and calmly responded "Doctor, do you know how much it cost me to have those put in?" This patient has a perfectly fine interpreter trying to make sense of what she knows and feels and does.[12]

The interpreter then is a like a detective who looks for clues and tries to come up with a theory that explains the facts. On the other hand, the interpreter is not the kind of detective we admire, because it gladly accepts the most threadbare evidence; "case closed, time for the next one," seems to be the operating policy of the mind's interpreter. There is no time to thoroughly investigate the cause, or more likely multiple causes, of an experience now arising in consciousness. The stream of consciousness usually moves too swiftly for thoroughness. New experiences will soon arise and the old will fade from working memory within half a minute unless heeded and recycled. Thus, the interpreter is like a very badly overworked detective who simply jumps on the first plausible theory that makes sense for the moment and moves on. Arriving at an immediate interpretation of consciousness, even if shaky, is apparently better than being at a loss for understanding.

The interpreter seeks explanations of our perceptions. It makes attributions about why events unfold as they do. It can be understood as the means by which the mind avoids uncertainty and maintains control. When viewed from this perspective, the interpreter should be especially active when faced with an acute feeling that matters are out of one's personal control. Then, the interpreter should especially seek to establish control by making connections, pulling together explanations, and seeking coherence even in the face of chaos.

A long line of research in cognitive science has documented that people make causal attributions about events as a means of maintaining personal control.[13] It is the feeling that things are spinning out of control that motivates the human brain to find a pattern in events and try to predict what is going to happen next. The left-brain interpreter thus will be activated whenever the individual senses a lack of control. Superstitions and conspiracy theories can be seen as the societal consequence of the interpreter's drive to find a causal explanation for events that are seemingly out of control. Jennifer Whitson and Adam Galinsky gave two such examples:

> Tribes of the Trobriand islands who fish in the deep sea, where sudden storms and unmapped waters are constant concerns, have far more rituals associated with fishing than do those who fish in shallow waters. . . . Baseball players create rituals in direct proportion to the capriciousness of their position (for

example, pitchers are particularly likely to see connections between the shirt they wear and success).[14]

In a laboratory study of personal control, the participants were shown a series of snowy, visually noisy pictures; half of the pictures contained a hidden figure and half contained nothing but noise. Half of the participants were first induced to feel a loss of control by trying to do a learning task in which they received phony feedback. Instead of being told when they responded correctly and when they did not, the feedback was entirely random. These participants in fact experienced a feeling of loss of personal control compared with those not receiving the phony feedback. It turned out that these vulnerable participants in fact were most likely to perceive meaningful figures in pictures that contained nothing but noise.[15] After experiencing a loss of control, these individuals struggled to regain control by interpreting nothing as something. Their left-brain interpreter made sense of the noise by seeing a hidden figure where none existed most often in the group experiencing a lack of control.

WHY THE LEFT HEMISPHERE?

What, then, are the parts of the cognitive machinery called the interpreter? There are two: causal inference and inner-directed language. Both of these parts are housed in the dominant left hemisphere of the brain. The interpreter's work, however, is supported by memory regions that are distributed widely in both cerebral hemispheres.

Causal Inference

The interpreter discovers or makes up, if needed, an explanation for why things happen the way they do. One of the specialties of the left hemisphere, then, must be the capacity to make inferences. The events that comprise the stream of conscious experience are reflected upon by the left hemisphere and explained by making assumptions about cause and effect. Event A is inferred to be the cause of Event B, based both on observation of their relationship in time and on knowledge of how the world works. For example, if we first observe a person shaking the branch of an apple tree and next observe an apple falling to the ground, then an inference is required to establish a causal relationship. The concept of force must be activated and seen as relevant to the observed facts. The force itself is hidden, but we likely infer that the shaking produced a force that caused the apple to fall. The force of gravity is also hidden, but from experience in the real world, we learn that unsupported objects fall to the ground. If the apple, dislodged from a branch, did not fall, then we would be very perplexed as to why it violated the scheme of our world knowledge. If the shaking did not dislodge the apple, then we would infer that the force exerted on the apple was insufficient, but we still would assume that a force was applied based on what we know from experience.

The causal inference just described is subtly different from the direct perception of causality. If instead of shaking the branch, the person directly strikes the apple and it falls, then no inference is required. We can visually perceive causality of one object affecting another. Direct perception of causality has been isolated as a specialty of the right hemisphere. Imagine a billiard ball—the white cue ball—rolls toward a stationary black eight ball. It strikes the eight ball, which then rolls in the same direction that the cue ball had been traveling. Using images on a computer screen, this scenario was presented to the left or the right hemisphere of a split-brain subject with the rolling ball labeled A and the struck ball labeled B. Inserting a time gap between A striking B and B starting to roll disrupts the perception of causation. Of interest, only the right hemisphere was sensitive to the experimenter delaying the movement of B.[16] The left hemisphere was not adept at direct causal perception.

Now, consider a different task that requires causal inference. Imagine two switches, one green and one red, positioned above a light. The subject first observes that depressing the green switch turns on the light. Second, he observes that depressing both switches also turns on the light. Third, he observes that depressing the red switch only does not turn on the light. Fourth, he again observes that depressing both the green and red switches turns on the light. Finally, a probe is presented and the subject predicts what will happen to the light. Either a red probe or a green probe is presented. To successfully predict that the green, but not the red, probe should turn on the light, the subject must make a causal inference. The causality cannot be perceived directly because a hidden force is involved related to the electrical wiring. The subject must reason that the light only comes on when the green switch is depressed, regardless of what happens with the red switch. The key finding with the split-brain patients shows that now the left hemisphere made the correct response to the two probes.[17] The right hemisphere was unequipped to predict accurately because it failed to make an inference of causation.

The capacity for causal inference could be by itself a uniquely human characteristic. Experiments with apes and monkeys have demonstrated their ability to learn through trial and error how to use a stick as a simple tool to obtain food from a clear tube. They eventually get the hang of using the stick to push the food out of the tube, and in that sense they have some understanding of the concept of physical force. However, by modifying the task slightly, limitations in their understanding emerge. In the modified version, a small trap is added under one part of the clear tube. If the primates "appreciate the causal force of gravity and the physics of holes and sticks moving objects, they should learn to avoid this trap as they attempt to push the food through the tube (i.e., they should always push the food out the end away from the trap)."[18] Neither capuchin monkeys nor chimpanzees get the hang of this version—it takes them dozens of trials to figure out avoiding the trap. By contrast, children only two or three years old "behave much more flexibly and adaptively on these tube problems—seeming to understand something of the causal principles at work—from the very earliest trials."[19]

Imagine you observe a person shaking a tree and seeing an apple fall. Now, suppose a strong gust of wind in a violent storm shakes the tree and an apple falls. Your mind sees these two situations as identical at the level of inferred causation. In fact, you can predict that a good way to get an apple to fall would be to shake the branch yourself. Remember that the chimpanzee's mind is organized around perceptual episodes rather than in terms of abstract concepts, such as force. To the chimpanzee, these three situations are three different episodes. In the view of Michael Tomasello, "Non-human primates understand many antecedent-consequent relations in the world, but they do not seem to understand causal forces as mediating these relations."[20] Knowing that a strong gust of wind can shake an apple loose does not prompt the chimpanzee to conclude that it might be a good idea to shake the branch on its own. Thus, the capacity for causal inference, as well as language, sets the interpreter of the human mind apart from even our most closely related species.

The visual and spatial stores of working memory support the inference capabilities of the interpreter. They provide temporary storage and replaying of perceived events. By holding in mind multiple events, hidden forces can be hypothesized and causal inferences reached. For example, holding in mind recollections of an apple falling from a tree when either the tree was shaken by the wind, by another person, or by the self, the inference that all of these events are linked becomes more likely. Working memory can also be used to imagine variations on past events. Once a causal inference is made that force must be applied to the tree, by whatever means, then the invention of new ways to bring down apples can be imagined in the mind's eye and simulated in working memory. A machine that vibrates the branches automatically could be conceived and then tested in the mind before actually trying it out for real.

The left hemisphere is specialized for drawing multiple kinds of inferences.[21] For example, when a split-brain patient is shown a picture of a pan and another picture of water, he can readily point with his right hand to a picture of boiling water. When this same kind of test is done with the right hemisphere and the left hand, the patient is unable to draw the simple inference that a pan and water go together to become a pan of boiling water. Similarly, syllogistic reasoning tasks also require inferences, and these recruit heavily the left hemisphere. For example, a verbal reasoning task might take the form of *all dogs are mammals; some mammals are large; therefore, some dogs are large.* Is this a valid conclusion? Neuroimaging of the left hemisphere reveals significant activity as the conclusion is evaluated. Even when a spatial reasoning task is presented instead of a language-based task, "the left still dominates in reasoning about those relations."[22] Recall that the right hemisphere is generally biased toward effectively processing visual-spatial information, so this involvement of the left in reasoning through the logic of spatial relations is telling.

Narrative and Language

The first tentative words of infancy turn into a torrent between the ages of eighteen and twenty-four months. This age range marks the shift from a single-word stage of language production to using two words, and it starts the child on a path to eventually uttering complete phrases and finally grammatical sentences. This skill in using language for communication with other human beings also eventually becomes the inner voice of the interpreter. However, in the early years, the child's interpreter speaks aloud in self-talk rather than silently. There is a slow transition in which the self-talk of the interpreter is vocalized just as if the child were speaking to her mother, brother, or friends. A young child in her preschool years can often be heard talking to herself aloud. The child is talking as if there were another person listening, but the speech is not intended for others. Rather, the speech is egocentric and intended for only the speaker to hear. As Lev Vygotsky concluded, "the child is thinking aloud, keeping up a running commentary, as it were, to whatever he may be doing."[23] This commentary eventually becomes interiorized around the time the child enters school, persisting as the inner voice of the interpreter.

Talking to ourselves aloud in early childhood is thus a transitional phase to the inner voice that accompanies us every waking moment for the remainder of our days. Perhaps egocentric speech is the fossilized remains of the speech capacity of early human beings when it first turned inward to the self rather than outward to others. Egocentric speech may be a reminder of how the ancestral human mind coopted speech as means of social communication and turned it inward as a medium of pure thought and a part of the interpreter.

Prior to the invention and adoption of language, was it possible that the causal inferences were made and a narrative constructed through mimetic images? Could the interpreter have once existed in the mimetic culture of *Homo erectus* without language? An affirmative answer is just speculation, but cognitive psychology can draw a distinction between the capacity for narrative thought and language. Narrative is a way of describing events that causally link together into a coherent whole—a story. Although narratives are often told using language, a story can also be portrayed with visual-spatial images.[24] Imagine, for example, a silent film. Or, an even simpler example is a cartoon, one without any of the characters saying anything in the little bubbles above their heads. Our dreams provide another example of the capacity for narrative thought in the absence of language. Although dreams can contain dialogue, language is not necessary for dreaming—the imagery by itself is sufficient and stands on its own as a story, albeit often a chaotic, incoherent one. Of course, mime is a prime way of telling a story without language. It is, therefore, conceivable that mimetic images combined with the ability to make causal inferences provided for a prelinguistic version of the brain's interpreter.

Nevertheless, the interpreter of the modern mind makes heavy use of narrative as expressed in language. Although the social benefits of language as a means for communicating with other human beings are unquestionable, the private, interiorized language of the interpreter is just as important to human nature. Subvocalized speech provides running commentary on the panoply of perceptions, memories, and fantasies that stream through consciousness. If we think of the ongoing conscious activity of wakeful cerebral hemispheres as a basketball game, then the interpreter is the voice of the play-by-play commentator.

Everyday thinking and daydreaming reveals a persistent inner voice, as shown in studies that interrupt daily activities with a beeper, followed by a self-report of the contents of consciousness. These reports indicate that about four distinct thoughts pass through awareness in working memory per minute, implying approximately four thousand distinct thoughts during the sixteen hours or so we are awake each day.[25] Self-talk accompanies these thoughts about 75 percent of the time; in 25 percent of the cases, the inner voice utters only a few words, but in the remaining 50 percent the internal chatter involves reasonable complete statements and running narratives. Eric Klinger illustrated the kinds of chatter reported by his participants as follows:

> The single most common feature of daydreams and other thoughts is self-talk. We hear an unexpected sound and say to ourselves, "What the heck was that?" We walk along appreciating the nice weather, and comment to ourselves, "What a nice day!" We play through our minds an image of a friend. We see him in our daydreams smiling and talking and we think in nearly so many words, "I wonder if he'll visit. I wonder if he really cares about me."[26]

Self-talk allows the mind to reflect upon its own experiences. It does so using all the sophistication of language to refer to a complex concept with merely a name. Through the syntax of language, one thought can become subordinated to another and long chains of thoughts can be concatenated. The inner voice makes possible complex patterns of thought that benefit human reasoning, problem solving, decision making, and the capacity to regulate our emotions and behavior. Language may have evolved as a biological adaptation. Improved communication among people allowed for greater cooperation and social cohesion of the group, which has clear advantages for reproduction and survival. But language turned inward in the form of the inner voice might well have been an *exaptation*, the term used by biologists to describe an adaptation that is coopted for a new use. Perhaps once language was established for communicating with one another we found the usefulness of self-talk.

The inner voice of the interpreter benefits from the sound-based form of verbal working memory. This store allows one to maintain thoughts in conscious awareness as the sounds of language, or at least a fragmented version of language. The words of inner speech are held in mind long enough for their meaning and syntactic structure to be grasped, providing a way for thought to gain coherence over time, just as it does in comprehending the outer speech of another person. Thinking in words and the running commentary of the interpreter would be impossible without the transient holding device of verbal working memory.

The inner voice of the interpreter can also be recycled in what can seem like an endless loop, as when a song's lyrics gets stuck in your head or when a voice from the past keeps intruding on an unwanted conversation. For example, think back to your own recent experience of an emotionally charged social exchange with someone close or with an antagonist at school or at work. How many times must a conversation be replayed in our heads before we arrive at an interpretation of why he said what he did last week or why she did what she did yesterday evening?

Our relationships with other people are often managed by attributing causes to our social experiences, and these causal inferences are often made on the basis of rehashed conversations. At the same time, the verbal store can be a vehicle for reworking these conversations (i.e., "If only I had said this when he said that," he thought). Past verbalizations can be reimagined just like past visual-spatial events. Working memory can even keep active in consciousness the causal inferences reached by the interpreter ("I can't believe how mean he is to have said that!"). The interpreter can thus ruminate about its own interpretations. All are grist for the interpreter's incessant milling of our conscious experience.

In sum, the interpreter of conscious experience is a left-brain mechanism that ingeniously combines the weighty powers of language and causal inference. Although each part is valuable, in concert they exceed the sum of the parts. The interpreter further draws upon long-term memory to provide a running commentary on consciousness. Working memory can prolong a perception, giving us time to make an interpretation. It can also prolong the interpretation itself as the mind ruminates and stews.

THE SELF AND THE INTERPRETER

Of the tens of thousands of concepts populating a human mind, one stands out as most critical for the entire enterprise. The self-concept is central in regulating our emotions, perceiving the actions of other people, and other cognitive functions. Psychologists theorize that the self is the *sine qua non* organizing structure of the mind. The storage of new events in long-term memory is facilitated by relating the event to the concept of one's self, for example.

The self-concept is a multidimensional entity that changes over time.[27] For example, personal identity includes the characteristics of our physical and psychological makeup, but also externally oriented roles that we play in social settings and our social status. It is formed in part by inspection of our own actions, but it is embellished, too, on the basis of what we learn from other people about ourselves. Social comparison to others provides a lens through which the interpreter can make inferences about the self. From very early ages, children compare themselves with others in terms of their height, their skill, their aggression, their friendliness, and so on. The interpreter takes in these social perceptions of how one stacks up relative to others and uses them to reach conclusions—for good or for ill—about the self.

That the self-concept is tightly knitted with the interpreter of the left hemisphere can be demonstrated through split-brain research.[28] The split-brain patient named J. W. was shown a set of nine faces. One of the nine was an image of his face. Another was an image of Michael Gazzaniga (M. G.). The remaining seven faces were computer-generated images that began looking very similar to J. W., but then gradually morphed into looking similar to M. G. Each step represented a 10 percent change in facial characteristics away from J. W. and toward M. G. The question is whether the left hemisphere would differ from the right hemisphere in its identification of the images as representing the self. If self-recognition is specially supported by the interpretive system of the left hemisphere, then seeing only 60 percent, or 50 percent, or maybe even less of the self in the image could prompt self-recognition. The results were startling. When the faces were shown to the left hemisphere of J. W., he recognized himself when the image contained only 40 percent of his own features. For presentations to the right hemisphere, it required twice as much relevant information—80 percent of his features—to reach self-recognition. In other words, the left hemisphere was far more likely to infer that the picture was a self-image.

The development of the self-concept does not emerge in a clearly recognizable form until between the ages of eighteen and twenty-four months.[29] A child of this age grasps the difference between self and other. This is a crucial development in the growth of the self-concept and the accumulation of a self-history. There must be an "I" and "me" to which events can be attached in long-term memory. Once the self is established, the young child can begin to accumulate new experiences over time, each becoming a fresh entry in the child's autobiography—a mental book of memory of and for the self.

Once the self-concept emerges, it begins its role as the central organizing cognitive structure. Prior to that time, memories are unstable and readily washed over by the daily waves of new experiences. Autobiographical recollections usually only begin around the ages of three or four years and never before the self-concept is established at twenty-four months.[30] The loss of very early memories during the first two years of life is known as infantile amnesia. Sigmund Freud famously attributed it to the repression of unpleasant, anxiety-provoking memories. Today, cognitive psychologists see it as an inevitable consequence of a mind not yet formed, a mind lacking the central organizing structure of a stable self-concept.

The self can change over time in some facets while remaining constant in others. The current contents of working memory at any given point in time provide the raw material for the interpreter to make attributions about the self. If some aspect of one's behavior is especially salient at the moment—such as avoiding interaction with others or feeling in a bad mood—then the left-hemisphere interpreter concludes that the self is shy or depressed, as the case may be. But such thoughts and feelings eventually fade from working memory—they always do, no matter how salient they may seem at the moment. The interpreter, presented with new evidence of the self as seeking social interaction or feeling elated, then concludes that the self is extroverted and happy. Thus, the self-concept changes shape over time as the interpreter accounts for the latest news to enter working memory. Although observing one's own behavior can be a source of inferences about the self, our internal thoughts, emotions, and motivations are especially salient.[31] Perhaps our actions are more likely to unfold mindlessly and so are less accessible to the interpreter. By contrast, our thoughts and feelings are by definition active in working memory and readily accessible for reflection and interference by the interpreter.

ILLUSORY INTERPRETATIONS

The brain's interpreter, the cognitive system that assigns the causes behind events and explains, moment to moment, an individual's experiences, has an interesting bias. In a normally functioning brain, thus excluding cases of severe psychopathology such as depression, the interpreter doggedly maintains a positive image of the self, as if it were viewing the world through rose-tinted glasses. In her book *Positive Illusions*, Shelley Taylor documented how people tend to regard themselves, the events of the world around them, and the events fading into the past as well as imagined in the future as more positive than the facts warrant. Normal human thought is, in a sense, taking place in a hall of distorting mirrors. The interpreter sees not reality, but reflected images that make us look better than we really are and that bolster our self-esteem.

In constructing a historical account of some episode in one's personal past, the interpreter slants explanations in the direction of self-enhancement. The self takes on the central and important role in the positive things that transpired and others receive the blame for the negatives. Take, for example, a married couple recollecting an event that led to a fight between them. The wife will recall what transpired quite differently than will the husband—both will invent a personal history that is more self-justifying than an unbiased third observer might relate. Recollections of the past, as with perceptions of the present, are filtered through positive illusions of self-enhancement.

Young children are especially prone to positive illusions about the self. Shelley Taylor captures the essence of youthful self-aggrandizement in the following passage:

> Before the exigencies of the world impinge upon the child's self-concept, the child is his or her own hero. With few exceptions, most children think very well of themselves. They believe they are capable at many tasks and abilities, including those they have never tried. They see themselves as popular. Most kindergarteners and first graders say they are at or near the top of the class. They have great expectations for future success. Moreover, these grandiose assessments are quite unresponsive to negative feedback, at least until approximately age seven. Children see themselves as successful on most tasks, even ones on which they have failed.[32]

By late childhood, our natural inclination to see ourselves as heroes destined to success has abated to a degree, but it never leaves us completely, even as adults. Part and parcel of the positive illusions of even adult thought is a conviction that we can control events to a far greater extent than is actually possible and that we will steer outcomes toward positive rather than negative outcomes. For example, most adults believe that they are less likely than others to be involved in a car accident because they believe that they possess above average driving skills. Obviously, we cannot all be above average, but this is what the psychologist must conclude from listening to the interpreter of each individual mind telling the story. Some people prefer the risk of driving a car on a long trip rather than flying on a commercial aircraft because they believe in their ability to prevent an accident through personal control. The reality is that traffic fatalities are far more common than commercial aircraft fatalities, from the massive numbers of cars on the road alone, and bad things can happen that are beyond personal control.

Maintaining a positive illusion of control can be thought of as highly adaptive in coping with inherently risky situations.[33] In the absence of convenient mass transit in some places in the industrialized world, driving a car is often the only realistic transportation option for trips too far to walk or bicycle. Believing that you are in control as the driver puts fear to the side and lets you get on with your life. Without this illusion of control, the alternative might be fear and paralysis.

In other cases of risk, however, illusions of personal control are less helpful. Take, for instance, a belief sometimes held by heavy gamblers that they can exert some influence over the roll of the dice, the fall of the roulette wheel, or the card turned up next in blackjack or poker. The craps shooter blowing on the dice, putting some body English into the throw, or getting the dice first kissed for good luck illustrates the underlying hope that control over chance is possible. Fortunes have been lost on such illusions.

One of the best illustrations of the interpreter's built-in positive bias can be seen in the way we think about the future. Again, assuming that the brain is fully healthy and not suffering from depression, the future constructed by the interpreter is generally bright. We carry with us a positive illusion of optimism.[34] Some of us are more optimistic than are others, but, taken as a whole, the human species thinks, quite irrationally, that the future is likely to be better than the present and certainly better than the past. Instead of using our past and present happiness and circumstances as realistic predictors of the future, the mind's interpreter does quite the opposite. This healthy illusion of unwarranted optimism is not seen in depressed people. In fact, the pessimism of depression is one of the cardinal symptoms of the disease. Usually mental illness involves losing contact with reality, but here, oddly, the depressed person may be more in touch with reality than the healthy optimist.[35]

It is important to clarify that positive illusions differ from the Freudian psychological mechanisms of self-defense known as repression and denial.[36] Repression forces anxiety-provoking thoughts out of the focus of attention. It protects the self, or what Freud called the ego, by keeping disturbing thoughts out of consciousness. In repression, a disturbing reality (e.g., witnessing or hearing of the death of a close friend or relative) is recognized and accepted but kept outside of conscious awareness as a means of self-protection. With denial, the disturbing reality is not accepted in the first place—the interpreter avoids harm to the self by refusing to acknowledge the death, explaining it all as a mix-up in communications or a case of mistaken identity. Thus, both repression and denial are psychic means of distorting reality in important ways. Positive illusions, on the other hand, are merely favorable interpretations of reality. They do not alter reality; they merely cast reality in the best possible light, one in which the self is in control and the future looks bright.

What happens to the self if the brain suddenly loses its interpretive capacity and inner voice? A remarkable portrait of exactly this situation comes from Jill Bolte Taylor, a neuroscientist who suffered and survived a major stroke. After suffering a massive hemorrhagic stroke in her left hemisphere, she eventually, after six years of heroic rehabilitation, recovered and went on to lecture and write about the day it happened. As a neuroscientist, she was able to describe with clarity and accuracy the experience of losing the critical regions of the left cerebral hemisphere that support causal inference, working memory, and language. When these became dysfunctional, her inner voice fell silent, stopping the familiar self-narrative that normally accompanies every moment of wakeful consciousness. What remained was a state of awareness mediated by her fully intact right hemisphere. She described the experience in her book *My Stroke of Insight*:

> I existed in some remote space that seemed to be far away from my normal information processing, and it was clear that the "I" whom I had grown up to be had not survived this neurological catastrophe. I understood that that Dr. Jill Bolte Taylor died that morning, and yet, with that said, who was left? . . . Without the language center telling me: "I am Dr. Jill Bolte Taylor. I am a neuroanatomist. I live at this address and can be reached at this phone number," I felt no obligation to be her anymore. . . . I stopped thinking in language and shifted to taking new pictures of what was going on in the present moment. I was not capable of deliberating about past or future-related ideas because those cells were incapacitated. All I could perceive was right here, right now, and it was beautiful.[37]

The left-hemisphere interpreter, then, is crucial to our inner sense of self. The interpreter shapes the memories of who we have been and who we will become, inscribing our autobiography in the pages of long-term memory. But how is it possible for the interpreter to travel back into the past and forward into the future? Just what is this remarkable ability made possible through what cognitive neuroscientists call episodic memory?

Chapter 6

MENTAL TIME TRAVEL

I n 1985, Michael J. Fox traveled backward in time to 1955 as the character Marty McFly in the movie *Back to the Future*. There Marty met his teenage mother, who at the time of his teleportation backward in time did not think much of his nerdy father, who was the laughingstock of the high school. Marty managed his miraculous time travel thanks to Doc, the mad scientist who invented the "flux capacitor" and mounted it on the dashboard of a DeLorean sports car. As Marty discovered, there were both ups and downs to traveling through time. What would happen, for example, if his then teenage mother developed a crush on Marty—her future son—and steered clear of his father completely? Would Marty simply disappear from existence, erased from history by a logical conundrum in the space-time continuum? Time travel, not surprisingly, is a staple of science-fiction literature, going back to H. G. Wells's classic *The Time Machine* in the late nineteenth century.

No less remarkable, or hazardous, is the human capacity for *mental* time travel. The mind can go back in time to relive past experiences or move forward in time to imagine future events. The cognitive system responsible for this mental version of time travel is called episodic memory. It is different from semantic memory in that the time and place of the experience is part of the memory. Its most unique feature is the capacity to turn back the arrow of time in the sense that the one doing the remembering is traveling backward into her past. Endel Tulving, who first defined and pioneered the study of episodic memory, articulated the marvel of this feat of the human mind:

With one singular exception, time's arrow is straight. Unidirectionality of time is one of nature's most fundamental laws. . . . The singular exception is provided by the human ability to remember past happenings. When one thinks today about what one did yesterday, time's arrow is bent into a loop. The rememberer has mentally traveled back into her past and thus violated the law of the irreversibility of the flow of time. She has not accomplished the feat in physical reality, of course, but rather in the reality of the mind, which, as everyone knows, is at least as important for human beings as is the physical reality.[1]

The capacity of episodic memory is crucially linked to the ability to travel forward in time as well as to travel backward. That is to say, one is able to imagine an experience in the future using the same brain/mind system responsible for remembering past experiences. To appreciate this point, it is important to understand that retrieval of the past event from long-term memory is not like retrieving a book from a library or a file from a computer disk. Instead, retrieval is an act of reconstructing the past experience and using one's creative capacities to fill in the gaps as needed. Reconstruction of the past is in principle really no different than constructing a future scenario. In practice, the past provides more guidance as to what fits in the reconstruction and what does not. Also, for most of us, the future is too much of a blank canvas and so we sketch in the outlines of a future scenario by recollecting a relevant past episode as the point of departure.

Long-term memory is a complex system that is organized hierarchically. At the top of the hierarchy, cognitive psychologists contrast declarative memory—storage of what X is—with nondeclarative or procedural memory—storage of how to do X.[2] Declarative memory divides into two distinct branches, episodic memory and semantic memory. The episodic part holds memories of specific past events and experiences while the semantic part is a storehouse of all our conceptual and factual knowledge. The semantic component of long-term memory allows one to know things about the world without remembering exactly how this knowledge was obtained. Semantic memory makes no reference to specific episodes situated in space and time. For example, suppose you spot a red sports car on the street. Recognizing the object as a car draws on semantic memory—your knowledge of the concept of cars, the superordinate concept of vehicles, and the subordinate concept of sports cars. Such memory allows you to list the properties of sports cars that distinguish them from other kinds of cars. It enables you to know that red is only one possible color for the car, and a rather dashing one at that. By contrast, if the sports car is one you have seen in the past—at a specific time and place—then that recollection is dependent on episodic memory. What, when, and where you experienced the object or event in the past is an inherent aspect of the memory. To take a different example, knowing what a birthday party is depends on semantic memory; recollecting your tenth birthday party requires episodic memory. In the laboratory, human beings routinely are able to distinguish between what they know to be true about information presented earlier in the experiment and the mental experience of traveling back in time and remembering its occurrence.

The two aspects of declarative memory are independent of each other, as dramatically revealed by the case of a patient identified by his initials, H. M. To control his severe epileptic seizures, surgeons removed most of his hippocampus from the limbic system on both the left and right side, along with some surrounding tissues in the temporal lobe, including the amygdala. The operation was a success from the standpoint of the seizures, but medicine was unaware at the time of the critical role that these brain regions play in episodic memory. It was discovered soon after the surgery that the procedure had resulted in a severe impairment of memory.

H. M.'s intelligence was not generally impaired by the surgery, which left his semantic store of concepts and facts about the world intact.[3] In fact, because the surgery successfully reduced the frequency of his epileptic seizures, his IQ score actually improved somewhat following the operation. However, it caused a profound case of anterograde amnesia that prevented H. M. from being able to store new experiences in long-term memory. This meant that H. M. could not remember events that had occurred a few minutes earlier. Once an experience faded from working memory, H. M. was unable to store and later retrieve that information on a more permanent basis in long-term memory. Thus, he lost the ability to store new events in episodic memory even though his semantic memory remained intact.

Although H. M. could not store new events in episodic memory, the autobiographical details of his preceding the surgery had already been stored successfully—these memories were not forgotten. He could easily tell stories about his days as a school boy, for example. Also not impaired were his semantic knowledge of concepts and his procedural memories for how to do things. He continued to mow the lawn as a household chore, to work jigsaw puzzles, and to read magazines. However, he had to be told "where to find the lawnmower, even when he ha[d] been using it only the day before." And he would "read the same magazines over and over again without ever finding their contents familiar."[4] Similarly, he could not remember the names of people whom he had met since the operation. Nor could he remember the location or address of a home he had moved into after the surgery. Instead, left on his own, H. M. would return to his old address a few blocks from the new one.

Another amnesia case important to understanding episodic memory is that of K. C., who suffered serious head injuries in a motorcycle accident that damaged many brain regions, including the inner or medial temporal lobes.[5] Tests on K. C. revealed normal intellectual skills. His working memory was also fully intact, meaning that he could remember new information for brief periods of time. There were no problems with his ability to read or to write, his ability to think clearly, or his ability to display previously learned skills such as playing the organ or chess. He could recall the facts of his life, such as his date of birth or the names of schools, but he was able to do so because they were objective facts stored in his semantic memory system. Similar to H. M., he suffered from a serious anterograde amnesia involving a deficit in storing new personal experiences, but at the same time K. C. exhibited a unique form of retrograde amnesia that was restricted to autobiographical experiences from the past that preceded his accident. K. C. was unable to remember any specific episodes even though he could remember factual information. That is to say, he could not travel back in time and recollect any particular personal experiences situated at a specific time and place, even when given various reminders about particular events. He denied any recollection or even any sense of familiarity with the particular event.

Thus, the consciousness of K. C. was cut off from the autobiographical past that human beings normally access through mental time travel. Although concepts and facts—the contents of semantic memory—were fully intact, he had lost access to an episodic-memory store that held the events of the past in relation to the self. Of equal importance and great significance, this absence of past memories was accompanied by an inability to imagine the future. "When asked, [K. C.] cannot tell the questioner what he is going to do later that day, or the day after, or at any time in the rest of his life. He cannot imagine his future any more than he can remember his past."[6]

It is difficult to say for certain, but it appears unlikely that nonhuman primates are capable of episodic recollection. Apes, and mammals in general, clearly are able to store knowledge about the world and use the hippocampus in learning new information. A vast scientific literature demonstrates clearly that the medial temporal lobe of the primate brain is critical to declarative memory. However, memory for factual information about the world is not the same as memory for specific events. As Endel Tulving has noted, "The kinds of tasks that have been used in evaluations of the hippocampal declarative memory system do not, and cannot, distinguish between memory for events and memory for facts."[7] Just as young children are able to learn facts without any difficulty before they begin to recollect specific autobiographical events from their personal past, apes may be fully capable of using semantic memory without any episodic memory. An ape may be fully capable of learning X but not have a recollection of when X was learned; the subjective experience of learning X, situated in the past, may be missing in all but human beings.

The thrust of most tests designed to assess animal memory makes use of where an event took place but rarely attempts to assess when the event occurred. The spatial but not the temporal aspect of episodic memory is typically addressed. Indeed, the essential feature of episodic memory is the subjective experience of mentally traveling backward or forward in time. Because the experience of another organism cannot be directly measured, and because apes cannot report their experience to us, this key aspect of time would seem to be outside the realm of investigation.[8]

One of the best attempts to demonstrate that nonhuman primates have at least the rudiments of episodic memory involved testing a lowland gorilla named King.[9] The gorilla was given food, such as an apple, at Time 1. Either five minutes later or as long as twenty-four hours later, King was rewarded for making a response that correctly identified what food he had been given earlier and who had given it to him. King was better at correctly identifying who gave the food than he was at identifying what the food was, but both dimensions were well above chance performance even after a twenty-four-hour retention interval. As the investigator pointed out, however, it is possible that King simply learned to store "the facts" of what was eaten and who provided the food in order to get a reward five minutes or twenty-four hours later. King was tested repeatedly in this setting, and he knew the test and the possibility of a reward would be forthcoming. It is also possible that King relied on the familiarity of "knowing" the correct food and provider without having a mental experience of remembering when the information was learned. The temporal dimension, especially the subjective experience of mentally traveling backward in time, cannot be readily assessed.

What is known for certain is that human beings are highly skilled mental time travelers. Our ability to think about the future may in fact be a crucial contribution to cultural evolution. Having an awareness of the future may be partly responsible for the human drive to innovate new and better ways of doing things—it is not just for us but also for the benefit of future generations. In other words, mental time travel may interact with our advanced executive function that enables planning and our advanced social intelligence that enables cooperation. Without an awareness of the future, nonhuman primates may "not initiate and persist in carrying out activities whose beneficial consequences will become apparent only in the future, at a time that does not yet exist."[10]

RECONSTRUCTIVE RETRIEVAL

The retrieval of an episodic memory is a reconstructive process. By understanding this fundamental fact about our recollections of the past, it is readily apparent how constructing the future makes use of the same cognitive machinery. Reconstructive retrieval is not at all like our everyday experiences of retrieving printed or digital information in our twenty-first-century culture. To retrieve a fact from a book, you find the book, open it, search for the fact, and read it verbatim; or you do the same thing within an electronic document. The same notion of search, find, and perceive applies also to video and audio information stored on compact disks or on the Internet. But with retrieval from episodic memory we draw on an ancestral form of memory for events. It is a form of memory that served us well in the oral culture of our prehistoric past, yet continues to do so today in a world of writing and audiovisual literacy, all supported by massive external memory via the Internet. The key difference is that in a purely oral culture, all information available to people *had* to be remembered using only the mind without the benefits of writing and other kinds of external systems of storage.

Reconstructive retrieval uses the concepts of semantic memory to create a plausible recollection rather than a verbatim recollection. Activated concepts create an expectation of what should have occurred in the past rather than what necessarily did occur. The concepts and facts of semantic memory are organized in mental structures called schemas. Schemas play a critical role in our perception of the world by providing expectations. For example, your ability to perceive the objects in the room in which you are now sitting is aided by anticipating what objects normally would appear in such a place. If you were sitting in a coffee shop a schema that organizes your past experiences with coffee shops would become active. The tables and chairs that you expect to see would differ from those you might expect in your kitchen or dining room of your home.

Related concepts are organized into a schema, which can be thought of as a cluster of closely connected concepts. The plenitude of schemas that make up all that we know about the world is, in turn, organized in semantic memory, again according to their relationships. The end result is a massively interconnected network of concepts and schemas. Activate one concept and there will be a path that the activation can travel to light up, eventually, any other imaginable concept. For example, think of a cat. This concept is closely linked to the concept of mouse. It is likely that you would then think of a cat prowling for mice. The concept of barn might then be activated, because it is a place where a cat might prowl for mice, but other concepts could also come to mind, such as house or backyard. But, if barn did come to mind, then the train of thought might immediately move on to the schema for a farm, with a cluster of related concepts popping into awareness (e.g., horses, pigs, chickens, cows, and crops). The network organization of semantic memory produces the associative structure of thought with which we are all intimately familiar. By following the associations through the network of semantic memory it is possible for virtually any thought, through some possibly lengthy chain of associations, to lead to any other thought.

Still, for any given schema and its associated concepts, some schemas are more closely linked than others. The farm schema is closely linked in semantic memory to the schema for the pristine countryside, for instance. By contrast, it is only very tangentially linked to a factory schema. By contrast, a factory is closely connected to the concept of city. It is difficult for people to think of farms as factories, because these ideas are so far apart in semantic memory. Similarly, the association of city and pollution is tightly wound in semantic memory, whereas farm and pollution are distant from each other. It tends to surprise us that laws are needed to prohibit pollution from large industrial farms that raise thousands of chickens or hogs. Whereas urban factories and pollution are immediately associated, farms are thought of as part of the wholesome countryside, not sources of toxic waste.

Our retrieval of past experiences from episodic memory passes through the organizational network of semantic memory. That is to say, these schemas provide expectations that help us to reconstruct past events. As part of their reconstruction, they can distort memories in order to conform to the expectations of the schemas active at the moment. Thus, schemas help us to reconstruct past events, but they do so by enabling us to fabricate how these events most likely unfolded. Reconstructive retrieval, then, refers to schema-guided construction of episodic memories that interpret, integrate, embellish, and otherwise alter the initial experience stored in episodic memory.[11]

A classic illustration of reconstructive retrieval comes from the downfall of America's thirty-seventh president, Richard M. Nixon. The testimony to Congress of John Dean, an assistant to the president, concerning the cover-up of a burglary at the Watergate Hotel, led to Nixon's resignation. Dean's testimony captivated Congress, the press, and the nation not only because of its factual content but also because of Dean's remarkably detailed recollections. Dean became known as the human tape recorder, given his confident recollection of conversations that had taken place weeks and months before. As it happened, Nixon had secret tapes of these same Oval Office conversations.

When the tapes became public, it was possible to compare Dean's actual conversations with his recollections, with the exception of a mysterious eighteen-minute gap in the tapes. Comparisons of the tapes with transcripts of Dean's testimony provided an extraordinary window on how episodic memory actually functions, not in a lab but in the Oval Office and Congress. Ulrich Neisser observed the following in his study of Dean's testimony:

> Comparison with the transcript shows that hardly a word of Dean's account is true. Nixon did not say any of the things attributed to him here: He didn't ask Dean to sit down, he didn't say Haldeman had kept him posted, . . . he didn't say anything about Liddy or the indictments. Nor had Dean himself said the things he later describes himself as saying.[12]

It was discovered that Dean's confidence in his recollections for details was unwarranted. Distortions showed that Dean hardly ever recalled the verbatim content of a given conversation, even though he was extremely confident that his recollection was veridical. In cases where his conversations were correctly recalled, it was more about the gist of the conversation rather than the details. Although Dean was truthful and accurate in conveying the gist of what happened, it turned out that "what seems to be specific in his memory actually depends on repeated episodes, rehearsed presentations, or overall impressions."[13]

Thus, Dean's recollections illustrate two important points about how schemas reconstruct episodic memories. The first is that the schema reconstructs the general idea or gist that is based on repetitions of essential elements or overall impressions. It is important to emphasize that Dean's testimony was factually accurate when considered at the level of gist. His recollections of the important facts were indeed correct, despite the small mistakes in the details. The second is that in recollecting an event, the schema richly embellishes it with details, as if it were a current event available for others to see and hear, without awareness that the details are fabricated. This is not an isolated example. Studies of eyewitness testimony have consistently found that the confidence of a witness is not predictive of the accuracy of recollection. One can be highly confident that a detail is correct, but still be dead wrong.

Thus, episodic memory is nothing like an audio-video disc that stores a verbatim record of a conversation or event. The record of human experiences can be altered at the time when information is initially encoded and stored in long-term memory. And it can also be changed in the process of reconstructing the event at the time of retrieval. How does the process of reconstructing an autobiographical experience unfold? For example, try to recollect your first day of school, your first kiss, your first day of your first job, your graduation from high school, the day of your wedding, or the birth of your first child. The process typically begins by retrieving the period of your life during which the event took place. People say to themselves, "when I lived in X" or "when I worked for Y." This is simply a fact available in the semantic store of long-term memory rather than a detailed episodic memory. Once a general period is identified, the retrieval process turns to themes and moods evoked by the event even though the specific event is not yet concretized. The theme and mood of falling in love, for example, sets the stage for remembering general rather than specific events. Finally, concrete images come to mind as sensory reconstructions of a specific event. Even though the phenomenal experience is of a recollection of the who, what, when, and where of a specific event, these episodic memories are always supported by and integrated with the themes and general periods of life that depend on factual or general information from semantic memory.[14]

A common trigger for recollecting our personal past is old photographs. Seeing oneself, family members, and friends in the places one used to live provides a retrieval cue for events from that period of life. In the process of reflecting on the picture and attempting to reconstruct long-forgotten events, the details can readily be mistakenly remembered, distorted, or totally fabricated without awareness of the error. For example, in one study, young adults were asked to recollect three events that took place at school during their childhood.[15] Based on a questionnaire completed earlier about their autobiographical experiences, two of these events were real, one that took place in grade five or six and another that went back a bit further to grade three or four. The third event, supposedly that took place in grade one or two, was actually false. This was an event that never actually happened, namely, the time a friend named Jane put slime in the teacher's desk. This pseudo-narrative was made up by the researchers and read to the participants along with accounts of the two real events. For each event, there was a group photograph of the student's class from the appropriate grades. Starting with the account from grades five and six, along with the group photo, the participants tried to recall as much as possible about the event and judged the extent to which they felt they were reliving the event. Finally, they judged their confidence that the event had really occurred all those years ago. Strikingly, seventy percent of those who looked at the photo and heard the description actually recollected the slime incident, with as much vividness as the real events. Without the photo, just under half remembered this false recollection.

Reconstructive retrieval recruits the interpreter in the left hemisphere. Like a detective attempting to reconstruct a crime scene, the interpreter does the same with autobiographical experiences from the past. Details are inferred, a storyline is concocted, and a coherent episodic memory can be constructed from whole cloth. The role of the interpreter in reconstructive retrieval has been highlighted with studies of split-brain patients.[16] In one study, a series of pictures was presented that portrayed a story. The pictures were shown either to the silent right hemisphere or the interpreter of the left hemisphere. In doing so, it was discovered that the right hemisphere remembers the literal story as portrayed only by the pictures. However, the left hemisphere made inferences that went beyond the literal pictures. These inferences allowed the left hemisphere to remember the gist of the story with various details embellished using inference and logic rather than any real recollection. Because the interpreter seeks to make inferences even when they may not be warranted, the left hemisphere plays a large role in false memories. From neuroimages collected as a person recollects past events, Michael Gazzaniga concluded that "although both hemispheres are activated when recalling true items, the left hemisphere becomes more activated when it is experiencing false reports."[17]

Before considering how the brain appears to accomplish mental time travel, it is instructive to experience reconstructive retrieval first hand. Try to remember what you were doing at this time of day one year ago. Take a few moments to think about the events that transpired one year ago at this time. When we try to remember an event from long-term memory, usually a schema from long-term memory is activated, one that is relevant to the circumstance in which you now find yourself and are trying to reconstruct. For example, suppose it is now 10 a.m. and you remember that one year ago you most likely would have been at work at that time. From this inference of where you would have been at such and such a time, the appropriate schema would become active, in this case a schema that summarizes the experiences from your working life, particularly those from the job you held a year ago. Or, if you instead inferred that a year ago at 10 a.m. you would have been in class, then a school- or college-related schema would jumpstart the recollection with expectations of what you were most likely doing at that time. With those expectations, reconstructive retrieval can begin to piece together and simulate an episode from the past. Notice that such a reconstructive process is much like solving a problem.

As another example, try to recall your tenth birthday party. Is your recollection accurate or is it, like John Dean's testimony, more of an amalgam of past experiences. Does it include experiences from birthday parties of friends rather than your own? Does it include the elements of past parties that occurred most often over the years of childhood rather than the specific elements of the tenth birthday party? Do your recollections include highly specific details that you were certain happened? Can you be certain, however, that they are not embellishments of the reconstructive process brought about by simulating the episode as if it were here and now?

Now, try to imagine what you will be doing at 10 a.m. one year from now. Is the process similar to recalling the past? Or imagine what your birthday party will be like ten years from now? The means by which the brain envisions the future is by recalling a similar situation from the past and modifying it appropriately. Cognitive neuroscientists have found that when envisioning the future, subjects reported situating the event in the context of familiar places, such as home, school, or work, and familiar people, such as friends and family.[18] In doing so, they were able to reactivate images from personal past experience and then modify them in ways that fit their expectations for the future.

Not only did their introspective reports reveal this phenomenon of constructing the future by reconstructing the past, but fMRI images of the brain confirmed it. The same neural regions that were activated in recalling the past appeared when the subjects were asked to imagine the future. These were the medial prefrontal cortex, the posterior cingulate cortex, the medial temporal cortex surrounding the hippocampus, and the occipital cortex in the left hemisphere.[19] In addition, an entirely new set of brain regions are activated in envisioning the future—these regions seem to be much the same as those observed when people are asked to simulate bodily movements. It seems as if the brain must actively simulate movements of the body in trying to form mental images of the future. In recalling events from the past, the bodily movements had actually once been experienced, so such imagined simulation is unnecessary.[20]

Of interest, this use of past episodic memories to see the future only occurs when the self is part of the memory. The self-concept and its close ally, the interpreter, are essential for the operation of episodic memory and its capacity for mentally traveling backward and forward through time. When the subjects were asked to imagine an event in the future with which they have no personal past experience, the results were different. They were asked to imagine not their own birthday in the future, but rather the birthday of a famous person, namely, Bill Clinton. The images generated in cases like this were less detailed and drew upon general knowledge stored in semantic memory rather than autobiographical knowledge. For example, subjects tended to report seeing Bill Clinton at a party in the White House, but the other guests were faceless and unknown to them.[21] This is indicative of drawing on a schema about birthdays from semantic memory but not engaging the interpretative self to construct the details needed for a true episodic experience of the future. Thus, when imagining one's future by traveling forward in time, episodic memory and the interpreter play a prominent role, but they do not when one imagines the future scenarios of a nonpersonal nature. The neuroimages of the brain confirmed this difference in reported mental experiences as well.

To summarize, episodic memory can be thought of as way of simulating the perception of an event in the here and now. As in the perception of an event at hand, schemas provide expectations about the contents and order of the past event, allowing a simulation of what must have transpired. Because episodic memory allows one to travel forward in time as well as into the past, the human capacity to imagine a future event is also a simulation. Because a future event has no record at all in long-term memory, the task of simulation is generally more challenging. One might start the imaginative process by reconstructing a similar event from the past and then modifying it as the mind projects into the future.

What areas of the brain become active when people are allowed to think about whatever they wish while neuroimaging of the brain takes place? Allowing the mind to wander as it will, instead of giving it a specific task to accomplish, reveals what cognitive neuroscientists call the default network of the brain.[22] When the brain is left to mind its own business, what does it think about? Of interest, it often turns to thoughts about the self, both reflections on past events and fantasies about possible future events. The default network thus involves the medial prefrontal cortex, the posterior cingulate cortex, and the medial temporal cortex. Besides remembering autobiographical events and envisioning the future, the default network is also activated in theory of mind tasks, such as when one must adopt the perspective of another person. It appears that the brain draws upon the default network to support "internal mentation that is largely detached form the external world . . . constructing dynamic mental simulations based on personal past experiences such as used during remembering, thinking about the future, and generally when imagining alternative perspectives and scenarios to the present."[23]

FALSE MEMORIES

Retrieval from long-term memory involves not only an effort to reconstruct the past, but also an attempt to monitor the source of the events that emerge from the reconstruction. Given that human memory uses the same mechanism for recollecting the past and for imagining the future, it is essential that we monitor whether the source of an experience is a result of past perception or of imagination. Without effective source monitoring, it is possible to confuse the two and believe that an imaginary event really happened. The reliability of episodic memory depends not only on having accurately stored information about one's past, but also on how well the information is retrieved and evaluated. The source of thoughts that come to mind must be monitored and attributed correctly to either reality or one's imagination.[24] For example, suppose you remember an embarrassing situation involving a particular person.

That person is someone you know and encounter in real life, and also someone who on occasion appears as a figure in your dreams. So, did the embarrassing event really happen or did it occur only in a dream? Source misattribution refers to false memories resulting from just such confusions. Effective reality monitoring, then, is an important component of sound episodic memory.

Psychologists have found that the more often people imagine an event occurring, the more likely they believe it has actually been perceived.[25] This again makes perfect sense given that our brain draws on common neural systems for perceiving an event in the present, reconstructing an event from the past, and imagining an event in the future. But it is problematic when the brain loses track of its time travel and mistakenly confuses imagination with perception. The more complex the event, the more perceptual detail there is to imagine and the easier it is to falsely remember it as actually having occurred. Those of us with especially vivid visual and auditory imagery are especially prone to these confusions of imagination with perception.[26]

Dramatic and highly consequential cases of false memory are part of contemporary society. Beginning in the 1970s and 1980s, allegations of childhood sexual abuse increased dramatically in frequency, including father-daughter incest.[27] These allegations were unusual in that the victims claimed to have no recollection of the abuse until years after it had taken place. In some cases, the memories did not surface until the individual was in psychotherapy for other current problems, such as depression, anxiety, substance abuse, or eating disorders. Therapists treating the individual regarded these recollections as recovered memories—recovered from a process of repression that had pushed the sexual abuse out of awareness for years. In fact, some therapists encouraged their clients to reconstruct the painful events through a variety of techniques, including the use of hypnosis. They further warned their clients that worries over whether the reconstructed memories were actually just fantasies and not true memories were normal but not to be trusted. Thus, the techniques simultaneously encouraged memory recovery and discouraged monitoring for false memories.[28] Encouragement to validate the recovered memories took the form of writing them down, vocalizing them to others in a group therapy session, or even publicly accusing the abuser.

It is unknown how many of these cases truly involved a repressed memory of sexual abuse. Because the sexual abuse of minors is known be perpetuated by pedophiles, some of these cases conceivably could have involved memory repression. However, it is also possible that the recovered memories reflected a failure of source monitoring. In reconstructing the events, it is possible that the products of imagination became confused with reality. Elizabeth Loftus and Katherine Ketcham conveyed the experience of several women under-going therapy to recover lost memories of abuse in what follows:

> At first the pictures in their minds included one abuser, usually a father, mother, or brother; but eventually the images enlarged to include uncles, aunts, cousins, grandparents, ministers, friends, neighbors. In the begin-ning, when the memory was developing, the abuse involved touching, fondling, probing; but as time went by the panorama of images expanded to include penetration, rape, and sodomy. Eventually, for several of these women, the mental spectacle included satanic cults, sadistic tortures, blood-drinking rituals, even murder.[29]

How is it possible to determine whether a recovered memory is true, partly true, or complete fantasy? Despite all the advances in cognitive neuroscience, there is no known method for answering this question. Cognitive science has methods for producing, measuring, and understanding false memories in the laboratory. Researchers have also worked with police and lawyers to aid in determining whether a person's testimony should be believed. However, "nobody has developed a neurophysiological procedure that can be used to predict whether a single memory is true or false."[30] Although this may be possible in the future with improved batteries of tests and neuroimaging techniques, it is not now possible.

However, the theory that recovered memories are commonly a result of repression is problematic for two reasons.[31] First, in some cases traumatic mem-ories may be stored in a highly fragmented form that is susceptible to being forgotten. Yet to the extent that fragments are forgotten for a period of time and then recovered, the recollections are prone to the distortions and inaccuracies that typify our normal mechanism of reconstructive retrieval. Second, in many cases, traumatic memories are not forgotten at all, but in fact are retained in vivid, relentlessly persistent detail. Severe physical and psychological trauma is

difficult to repress—consciously or unconsciously—despite the fact that forgetting would bring healing and relief. The repeated, intrusive, automatic flashbacks to severely traumatic events constitute a psychological disorder known as
post-traumatic stress disorder (PTSD). Sexual assault is known to be a cause of
PTSD, as is exposure to combat in war. Known as shell shock in World War I,
combat fatigue in World War II, and now PTSD in more recent wars, all refer
to the same problem: persistent, unwanted, and disturbing replays of death and
destruction, deprivation and despair. Sufferers of PTSD recall the trauma of
sexual assault and combat all too easily and frequently.

Both of these problems can be found in the most famous case of repressed
memory. At the age of twenty-nine, Eileen Franklin accused her father of being
a murderer. Twenty years earlier, her best friend, Susan Nason, had in fact been
murdered, and the case had gone unsolved. Eileen had a flashback in adulthood
that her father, George, was the perpetrator all those many years ago. Elizabeth
Loftus served as an expert witness at George Franklin's trial to provide testimony about the possibility that Eileen's recovered memory was not a consequence of repression but instead reflected a false memory. In her analysis of the
case, there were two arguments against the recovered memory theory:

> If stress (and, of course, time) cause memory to decay and deteriorate, why
> did Eileen Franklin's memory come back to her twenty years later in such
> astonishing, full color detail? If . . . traumatic events create clear, detailed,
> and long-lasting memories . . . , how was Eileen able to push the memory of
> Susan Nason's murder out of her conscious mind for nearly twenty years?[32]

Sexual abuse and murder are plausible events, because regrettably they do
indeed happen. But what about recovered memories of events that seem highly
implausible, more from the realms of fiction and film than from reality? The
tabloid press at times reports cases of seemingly sane individuals who recall
being abducted by aliens and taken aboard unidentified flying objects (UFOs),
being sexually abused during satanic rituals, or witnessing cannibalism of
children. From a scientific viewpoint, such reports are difficult to accept as
anything other than false because they are so bizarre and implausible. How,
then, might they be understood? One plausible explanation is that such

recollections are false memories instilled through a sociocultural mechanism of memory implantation.[33]

Through repeated suggestion from the information channels of the culture, beliefs about the world take hold. Through what a person reads, watches on television, hears on the radio, and discusses with others who receive the same inputs from the culture, a view of the world can take shape that has very little connection with objective reality. In the laboratory, cognitive psychologists have shown that "exposing people to a set of articles that describe a relatively implausible phenomenon, like witnessing a possession, made people believe that the phenomenon is more plausible, and also made them less confident that they had not experienced the event in childhood."[34] Such memory implantation is perhaps rare but possible if three conditions are met. People must first believe that an event is plausible. Second, they must undergo a process of falsely reconstructing an autobiographical experience that actually happened to them. Third, they must interpret their subsequent thoughts and fantasies related to the event as real memories.

Delusional false memories require more than suggestion and false beliefs, however. They further require an experience that arises internally but is misattributed to an external source. The mind becomes confused as to what is real and what is not. Just as in perception it might be difficult to tell the difference between a real object and a mirror image of an object, the same confusion can arise in judging the source of memorial experiences.

For example, if an event occurred in a dream, it is important to know when later recalling the dream that the event was not real. But this source information can be forgotten over time. The dream itself is wholly determined by the internal forces of past episodes and past beliefs. The interpreter attempts to weave together hallucinated perceptual experiences into a narrative, albeit often a bizarre and incoherent one. A delusional false memory might then occur if one forgets that the bizarre event was only a dream. Misattribution of source is most likely for individuals with strong beliefs that a bizarre event could actually occur. The social circles of the individual support and find entirely credible the narrative reports of a person experiencing a delusional false memory. There is no check on the misattribution of source, but only affirmation.

Consider stories about encounters with a UFO, for example. Betty Hill was an avid believer in UFOs before the night she and her husband, Barney, claimed to have seen a strange light following their car. Betty's sister suggested that she and her husband may have been "irradiated" by the light, and soon Betty began to have nightmares in which she and Barney were abducted by aliens who took them on board a UFO, communicated with them telepathically, performed medical tests, and showed Betty a star map. A psychotherapist who treated Betty for her nightmares used hypnosis to elicit detailed reports of these events from both her and Barney, who had heard these accounts from his wife numerous times. The therapist concluded that the reports were delusional false memories shared by the couple and were nothing more than fantasies.[35]

THE HAZARDS OF TIME TRAVEL

The phenomenon of false memories—whether they are plausible or implausible—highlight for us that episodic memory can operate both in the past and in the future. The reconstructive processes that allow us to remember the gist of an actual past life episode can also encourage us to fabricate one that never happened. The ability to envision future events—when turned backward in time—can result in confabulating a false past. Absent effective reality monitoring, the mind cannot discern factual events from fictional events. In this sense, false memories represent a hazard of mental time travel. It is not the only one.

Mental time travel gives human beings enormous power in the ability to recall the past and to plan for the future. Civilization itself is a result of the awesome capacity to envision how things could be in the future and the ability to recollect and learn from the past. But the escape that mental time travel provides from the present can become a disability rather than an adaptive cognitive tool. Clinical depression is characterized by the arrow of time persistently pointing backwards in rumination about the past, with life experiences being repetitively replayed and repented. Anxiety disorders, on the other hand, can be understood as a preoccupation with the future, not in joyful anticipation, but in worry and dread. Brooding over the future as well as repenting of the past can be pathological. Anxiety and depression can also co-occur, when one alternates between anxiety about the future and worrying about the past. The very capacity that frees human beings from the here and now can also trap us in a nightmarish past, a frightening future, or even both.

Imagining bad things that could happen in the future, anticipating situations and events that more often than not never occur in reality, is the hallmark of debilitating anxiety. Anxiety is difficult to define, but as human beings we know it when we feel it. It is something less specific than simple fear of an immediate danger. Fear can be a friend when it prepares a person to take action, to flee or to fight. Anxiety, by contrast, is not proportional to the threats of the immediate environment. In fact, it is not driven by the present situation at all, but rather by the mind's anticipation of future problems. Some anxiety is normal and beneficial when kept constrained, because it allows for planning and preparation to meet future demands. The problem arises when one constantly broods over the future and anticipates only negative events. Then, the machinery of imagination that can be so useful in planning for the future is turned against us as it persistently conjures dangers that will rarely, if ever, occur.

Imagining that the worst will happen naturally spurs negative emotional thoughts or worry. Worry is known to be highly verbal rather than visual-spatial in nature; individuals who worry excessively report an abundance of negative self-talk rather than unpleasant visual images.[36] It is expressed through inner speech and reverberated in verbal working memory. As such, worry is a product of an agitated interpreter fretting about the contents of conscious experience. Thus, those who suffer from anxiety disorders can be understood as having an overly active episodic memory focused on future problems. Through mental time travel, negative future events are fabricated and given context as visual-spatial images. Actual past events can serve as a starting point for these fabrications as they are embellished with troubling details and imbued with darker emotional tones. What had been an actual mild annoyance in the past can be envisioned as a calamity in the future. The interpreter, then, provides a narrative commentary of inner speech to accompany the gloomy visions of future miseries and possible catastrophes. The chattering of the interpreter only worsens the outlook through verbal worry.

We are an itinerant species, not only in our wanderings through the past and future, but also in our spatial movements all over the planet. Human beings have traveled from their place of birth since the origin of the species. Our migrations, through thousands of generations, from out of Africa to all corners of the globe, document the essential itinerant nature of *Homo sapiens*. In Homer's story of the wanderings of Odysseus, we see the consequence of our restless history. In the face of monsters and even the pleasures of the sea nymph Calypso, the mind of Odysseus persistently turns its arrow of time back into the past, as he longs to return home to his native island of Ithaca, where his wife Penelope faithfully waits for him. Homesickness is but one of many ways that nostalgia seizes the mind, however. *Nostalgia* refers to any recollection of autobiographical experience stored in episodic memory that attaches to an emotional wish to be there again in that time and place. Although returning home is one source of nostalgia, it can refer to people, events, and locales from other parts of life as well.

Psychologists have asked people to write a narrative of a nostalgic experience and then analyzed the content.[37] Family members, friends, romantic partners, birthdays, vacations, and specific times and places of emotional significance, such as observing a special sunset, were the most common objects of reverie. Nostalgia is in fact a common emotional experience associated with mentally traveling back into one's past. Besides being found across different ethnic and cultural populations, it occurs as often as weekly in the majority of people.

Although homesickness is usually associated with sadness, researchers have learned that nostalgic recollections more often than not evoke the positive emotion of happiness rather than sadness.[38] Also, some recollections of nostalgic events paradoxically give rise to feelings of both sadness and happiness. Such bittersweet narratives share a common structure, beginning with a disappointment or loss that elicits sadness, grief, or suffering, before moving forward to happy acceptance, triumph, and euphoria. In other words, nostalgic recollections seem to be a vehicle for redemption, a means of reliving negative emotional episodes from the past, coming to terms with them, and emerging from the reverie feeling better. It is particularly interesting that feelings of loneliness and, more broadly, negative moods in general, are often the triggers for turning the arrow of time backward. Episodic memory, then, has a role in our mental health by grinding down psychological discomfort into equanimity.

Yet the mental time travel of episodic memory can harm mental health in depression. While nostalgia may provide us with redemption, meaning, and purpose in life, rumination can eviscerate us through the mental anguish of regret. Episodic memory provides us with a window on our past that we look through with some peril. It provides the interpreter of conscious experience an opportunity to rework personal history and imagine what might have been. It is through looking back that regret for past choices breeds and multiples. Regret can, of course, be a healthy emotion, such as when it is part of a confession of past mistakes and when it motivates repentance and the resolve to behave differently in the future. However, past wrong doings are not the only objects of rumination. Events from the past over which one had little, if any, control can haunt the human mind. For personalities prone to self-criticism and poor esteem, for one whose interpreter always finds fault for the past within the self, rumination becomes part of pathological anxiety and depression. Rumination is also associated with binge eating, binge drinking, and self-harm.[39]

For a person diagnosed with clinical depression, the arrow of time points backward persistently. Hope for the future requires the capacity to even think about the future. Yet in the clinically depressed patient, negative memories from the past spontaneously come to mind, and the person comes to believe that most of his or her life has been little more than endless succession of painful events. The negative mood of the present seems to grease the retrieval of the negative past, while the positive experiences are forgotten. Worse, rumination on these erupting painful memories further depresses the individual's emotional state, causing a malevolent spiral downward. In severe cases, the embattled mind of the clinically depressed can even come to prefer the self-destruction of suicide.

For most people, severe depression is rarely or never experienced. Although depression has been called the common cold of mental illness, and periods of negative moods and sadness are indeed a fact of human experience, most people escape the depth of depression that motivates its victims to take their own lives. On the contrary, despite life's many assaults on human emotional stability—ranging from the daily hassles of working and surviving to the tragic deaths, divorces, bankruptcies, fires, hurricanes, tornados, earthquakes, assaults, and wars that sometimes befall us—people largely remain positive, optimistic, and resilient. How is it that such negative emotions can be so successfully regulated? We turn next to the limbic brain regions that emanate emotional experiences, and to the ways in which the neocortical parts of the modern ensemble modulate these feelings.

Chapter 7

EMOTIONS

Interest, joy, fear, anger, disgust, and sadness are basic emotions that human beings express in a species-specific manner. The facial expressions of fear, sadness, and joy, for example, are immediately and accurately recognized regardless of the culture and language of the person experiencing and perceiving the emotions. The nonverbal expressions of emotion are behaviors that are coded into our human repertoire just as is walking on two legs. But the basic emotions are only a part of our repertoire. Variants and blends of these also color human consciousness, such as guilt, regret, jealousy, envy, pride, shame, hate, and love.

An emotion may enter into conscious awareness in the form of images, as the end result of a cascade of responses that begin in the brain, spread throughout the body, and are sensed and perceived by the brain.[1] For example, a feeling of fear is initiated in the amygdala, which is buried deep in the limbic system of the brain. The system includes the hypothalamus, whose job is to bring about hormonal changes in the body. It does so by stimulating the pituitary gland, the overseer of a host of glands that together make up the endocrine system. The adrenal glands that lie on the kidneys spew out a stress hormone called cortisol that helps to raise the energy sources needed for immediate strenuous action. The hypothalamic-pituitary-adrenal cascade—known as the HPA axis—is one way that the emotion of fear manifests itself in the body. Another way is via the autonomic nervous system. Specifically, the sympathetic branch of the autonomic nervous system accelerates the heartbeat, suppresses digestion, stimulates the release of glucose into the blood stream, and releases epinephrine or adrenalin from the adrenal glands. The muscles are primed for a fight-or-flight response. Feelings of fear can then be consciously represented within the brain. Such feelings are thus similar to the auditory images of sound or the visual-spatial images of sight, but they arise into consciousness from perceptions of the body, whether they are the tension of the musculature, the pounding of the heart, or the knot in the stomach.

The biological function of emotions is to participate in the regulation of our behavior. Emotional feelings help determine whether to avoid a situation out of fear or approach it out of interest.[2] They thus help us to survive in much the same way that pain and pleasure do. The objects, events, and autobiographical experiences of our lives are linked with feelings of happiness, sadness, anger, and so on. For example, if an event elicits fear, the increase in heart rate and blood pressure that is part of the emotion facilitates a flight response to avoid the danger and threat to survival.

However, emotions—when not fully recognized and managed properly—can sometimes distort the outcome of thinking in the direction of irrationality, or worse, join forces with destructive impulses and thoughts. Carl Izard, a leader in the science of emotions, noted that

> unfortunately, linking emotion feelings to maladaptive thoughts like those that characterize racism, sexism, ageism, unbridled profit motive, and plans for vengeance, revenge, or terrorism can wreak extensive havoc to individuals, ethnic groups, and all of human kind. For an abundance of evidence supporting the foregoing assertion, read history and watch or listen to any daily news program.[3]

THE ROOTS OF EMOTION

Emotion is not unique to human beings. Mammalian species, besides giving live birth, share common structures in the limbic brain that play an important part in emotions. Lying below the neocortex and above and around the brain stem and primitive forebrain regions, the limbic system provides a foundation for the use of emotion to regulate behavior. The basic life support responses of the brain stem can be viewed as reflexive, automatic means of survival, such as regulating the internal environment of the body to maintain homeostasis. Through the limbic responses, the organism gains more complex means of regulating behavior through the feelings of fear, happiness, and other emotional states. For example, fear provides the signal for knowing what to approach and what to flee from—a regulation of behavior of immense importance for an individual's survival.

Despite the commonality and rudimentary nature of emotion, human emotion cannot be understood without an appreciation for the way the cognitive processes of the neocortex are laid over upon and modulate the responses of the limbic brain. In particular, the ensemble of the modern human mind shapes emotion in ways that render it different from the emotions of other mammals and even other primates. The advanced working memory of human beings is capable of holding in mind thoughts that can ramp up or dampen down an emotional response. The situation that gives rise to an emotional response can be cognitively appraised, and the outcome of this appraisal modulates the intensity of the feelings triggered. The interpreter makes causal inferences about the source of emotional arousal just as it does about other consciously perceived experiences. Mistaken attributions may not only diminish the intensity of an emotion, but even alter the basic emotion felt—what might have inspired fear can be experienced as exciting and attractive. The limbic circuitry for fear—perhaps the most adaptive emotion for an individual's biological survival—is now well documented by neuroscientists. Yet it is also known that the recycling of fearful thoughts in human verbal working memory can greatly exacerbate apprehension beyond any adaptive value. Phobias and generalized anxiety disorders are human afflictions enabled by the very power of our working memory. Similarly, mental time travel can strand one in the past, ruminating endlessly about negative emotional experiences that only serve to drive the afflicted deeper into depression. After experiences of severe trauma, mental time travel can so vividly capture the emotional horror in a flashback as to make the victim feel victimized all over again. All of these serve as examples of how the ensemble of the modern mind has altered the roots of emotion.

Richard Lazarus has explained that the limbic forebrain is really only the beginning of understanding human emotion. Such structures "provide a neural template that makes emotion in all species similar in fundamental ways."[4] However, he goes on to note that there is nothing in this perspective that requires the reduction of all emotion to the lowest common denominator of comparatively simple animals and reptilian and mammalian brain structures. When such reduction occurs, it is at the expense of recognizing and investigating the primary role of cognition in emotion. It is about time that we began to formulate the rules about how cognitive processes generate, influence, and shape the emotional response in every species that reacts with emotion, in every social group sharing values, commitments, and beliefs, and in every individual member of the human species.[5]

FEAR AS A PRIMAL EMOTION

Perhaps the emotion best understood by science is fear. Although its valence is clearly negative, fear is a beneficial emotion in motivating avoidance of danger. Its adaptive value in keeping organisms alive and able to reproduce is likely the reason why it emerged so early in the evolution of life. Fear can be observed throughout the animal kingdom, from fruit flies to people. Reptiles, fish, and all mammals exhibit a phenomenon known as fear conditioning, which has provided neuroscientists with a window into the specific neural circuits that mediate the bodily response and subjective feeling of fear.

A laboratory rat is first played a sound that has little if any impact on its blood pressure or its tendency to freeze in place, ceasing movement until danger has passed. This sound then is presented as a conditioned stimulus (CS) that precedes a painful shock, the unconditioned stimulus (UCS) that automatically generates fear, as can be seen in a sharp spike in blood pressure and several seconds of freezing in place. After several of these pairings of CS and USC, classical fear conditioning results in the same physiological changes when only the sound or CS is presented by itself. This animal model of fear conditioning has provided a way to determine the specific regions of the brain involved in the experience of fear.

An understanding of fear requires an integrated view of the limbic system and other parts of the brain.[6] Although the limbic structures of the amygdala and the hippocampus are essential parts of fear conditioning, the diencephalon and telencephalon also play a role. Specifically, the sound follows a pathway from the ear via the auditory nerve to the midbrain then to the thalamus and finally to the auditory neocortex. The output from the thalamus also goes to a cluster of neurons in the amygdala of the limbic system called the lateral nucleus. The signals from the auditory thalamus to the lateral nucleus offer a high-speed connection to the initiation of fear and the freezing of movement. Although the thalamus cannot process the identity of the stimulus that launched a fear response, it can quickly, in a single step, link to the critical structure of the amygdala.[7]

At the same time, a parallel computation is underway in the neocortex, where detailed and accurate representations of the external world are constructed for conscious perception.[8] This involves connections from the auditory thalamus to the auditory neocortex and then back to the amygdala. Many neural links are involved in this neocortical pathway, and so the registration of fear in the amygdala takes more time than via the subcortical pathway. Imagine walking in the woods in autumn and seeing a curved object lying in the leaves and hearing the soft sound of leaves rustling. Before we are able to identify the object causing the sound as a snake, the subcortical circuit has already jacked up our blood pressure and frozen our movements or prompted a jump backward away from the threatening sound.

Although the amygdala is essential to fear conditioning, another kind of learning is also taking place. Episodic memory—the capacity to store and later retrieve into consciousness a recollection of an event from the past—is consolidated by the hippocampus and stored in the neocortex. Specific episodes from the past and the factual content of our knowledge about the world make up the declarative memory system, which operates independently of the nondeclarative system that includes fear conditioning. Damage to the hippocampus interferes with one's ability to recall a past traumatic event, but the emotional learning of fear conditioning proceeds just the same outside of conscious awareness. On the other hand, when the declarative system is intact, people store memories of past trauma not as hot emotions but as cold facts. Joseph LeDoux distinguished the two in the following example:

> If a person is injured in an auto accident in which the horn gets stuck, he or she may later react when hearing the blare of car horns. The person may remember the details of the accident, such as where and when it occurred

and who was involved. These are declarative memories that are dependent on the hippocampus. The individual may also become tense, anxious, and depressed as the emotional memory is reactivated through the amygdalic system. The declarative system has stored the emotional content of the experience, but it has done so as a fact.[9]

The parallel pathways for fear conditioning, on the one hand, and the formation of declarative memories in the form of facts and specific episodes from our life experience, on the other, illustrate how complex even the primal experience of fear can be in the mammalian brain. The limbic system interacts with the neocortical regions even in rats, and such interactions are especially important in the human brain. The ensemble of parts that constitute the modern human mind can markedly alter emotional responses. The result is that the emotional life of a human being is not easily reduced to the output of the primitive limbic forebrain.

For example, in the mental disorders of generalized anxiety, phobia, and panic attack, the powerful human system of episodic memory is turned against the sufferer. In these afflictions, the healthy fear response that keeps us out of harm's way mushrooms into a toxically maladaptive and debilitating emotion. For phobics, emotional learning leads to irrational fear of highly specific situations, such as the claustrophobic fear of enclosed spaces. Fear may manifest itself also as generalized anxiety during which one may be unable to pinpoint exactly what objects or events are provoking the sense of apprehension. The intensity of these responses can vary. In the most extreme responses— called panic attacks—there is a four-alarm fire of emotional upheaval. Hyperventilation and hot flashes, feelings of intense anxiety and nausea, and the sensations of choking and fainting can overtake the victim in the throes of a panic attack. Agoraphobics' fear of open, public spaces is compounded by the dread that they may suffer a panic attack in public, where escape without notice is difficult. Seeking safe havens, sufferers of agoraphobia often confine themselves to their homes. In these and other examples, the maladaptive side of fear is all too evident.

In all anxiety disorders, the capacity to travel back in time and bring fear-provoking past experiences to mind compounds the problem. A deep-seated fear of heights, for example, may bring to mind a past episode in which one's life felt threatened by being on the top floor or roof of a tall building, or at the top of a roller coaster, or in the middle of a high bridge. A past bad experience can be replayed over and over again, deepening the fear, even though the fear-provoking situation is not immediately at hand. Our ability to travel back in time and recollect such fearful experiences using the default network of the neocortex becomes a way of triggering the fear reactions embedded in the networks of the limbic system. It is known that the medial temporal lobe that supports episodic memory sends projections back to the amygdala, and these could well trigger fear responses.[10]

Remembering past stressful episodes is only half the story, however, Mental time travel also enables the human mind to imagine the future, including mental scenarios of impending doom. The phobic might imagine that it is likely the feared object or event is lurking just ahead in their future. Efforts to avoid the encounter are then seen as essential and comforting. For example, the acrophobic might anticipate what high places might be encountered during the course of the day and think through how to avoid them. The use of mental time travel is especially pervasive in generalized anxiety, where fear is not restricted to a specific phobia. The ability to imagine scenarios of future problems is unleashed and the constant anticipating and dreading the worst takes its toll.

Besides episodic memory interplaying with the limbic system, a similar interaction occurs with advanced working memory of human beings and the interpreter of consciousness. As future threats are brought into mind in those suffering from severe anxiety, they are held there tenaciously and recycled through the verbal store of working memory. The inner voice of the interpreter narrates excessively about the bleak events of the future. Interpretations are assigned to these thoughts that only make them more persistent and more out of control. Those who worry excessively are familiar with these ways in which the advanced capability of human working memory and the capacity to carry on an inner dialogue can make life miserable. They become highly preoccupied with the negative "self-talk."[11]

The anxiety disorders reflect breakdowns in our ability to regulate the emotions generated by the human brain. They illustrate that the unique capacities of the modern human mind are not entirely beneficent. The neo-cortical systems that provide for advanced working memory, the interpreter of consciousness, and mental time travel can interact with the limbic structures of the brain in a way that harms psychological health. An even more striking example of this linkage can be seen when human beings are traumatized.

Post-traumatic stress disorder (PTSD) represents the residual effects of trauma. It is an anxiety disorder that can develop after a person has been exposed to the threat or actual experience of severe physical harm. Violent personal assaults in domestic or public settings, military combat, life-threatening accidents, and natural disasters, then, are among the traumatic events that can cause PTSD.

Intrusive thoughts and images of the trauma—even vividly recollected flashbacks—plague the victim. Feelings of emotional numbness and detachment are common, as if the mind seeks some relief in turning off the circuitry of emotional responses altogether. Sadly, the emotional numbness is often most severe when around the people with whom the sufferer had once been closest. The places, people, and thoughts associated with the trauma are avoided as much as possible. Signs of constant fear and apprehension are evident in the victim's state of irritability and hyperarousal. The sufferer feels jumpy and easily startled. Sleep provides no relief in that nightmares disturb slumber or insomnia prevents sleep in the first place. In PTSD, normal emotional regulation breaks down in the white waters of relived trauma much as a boat flounders in a hurricane. The capacity to relive a traumatic event as if it were happening all over again is the root cause of PTSD. While normally our human ability to recollect the past is a blessing, after the experience of severe trauma it can become a curse.

Experiencing or witnessing accidents, fires, tornadoes, earthquakes, or other natural catastrophes, or, even worse, traumas inflicted by human design, profoundly affects human emotions. Sexual molestation, rape, physical assault, and being threatened with a weapon all leave deep emotional scars in the mind of the victim. For those who identified rape as their most traumatic experience, 65 percent of men and 46 percent of women suffered PTSD as a result.[12] Similarly, high rates of rates of PTSD followed childhood abuse for women (48 percent) and combat exposure for men (38 percent).

The threat to one's life in combat is an obvious form of trauma. But there is more. Witnessing the atrocities of war—seeing unknown enemies as well as close friends maimed, murdered, or massacred—goes far beyond just fearing for one's life. It exemplifies a total collapse of the social norms of civilization, including the sanctity of human life and the bonds of shared humanity. PTSD, as a concept and a label, was not formulated until after America's war in Vietnam, but the same symptoms were observed and identified as shell shock in World War I and as battle fatigue in World War II. Besides the unbidden flashbacks of trauma during waking hours, the reliving of traumatic experiences also haunts the dreams of those who suffer from PTSD. Combat veterans who suffer from PTSD are frequently plagued by frightening and disturbing nightmares that reoccur on a regular basis. They can persist for decades.

A US Air Force officer, Michael Gold, had flown numerous combat missions in a B-17 over Nazi Germany before being shot down on January 30, 1944.[13] He was captured and spent the remainder of the war, more than a year, as a prisoner of war. After coming home, he went to college on the GI bill and then was accepted to the Rochester School of Medicine. He became a doctor and had a successful medical career. However, for decades after the war he suffered from recurrent nightmares about his war experiences.[14] Here is a recounting of one of them:

> He is trapped in the nose of a B-17, scrambling on his hands and knees. The plane is on fire, wallowing through the clouds, going down. Someone is screaming. His lungs fill with smoke; everywhere the smell of cordite and burning rubber. Spent shell casings clatter beneath him. Wind howls through the plane. He struggles toward the escape hatch. Crawling, crawling in the numbing cold. The hatch slips away, receding farther and farther, fading from view. Vanishes.[15]

The nightmare replayed elements of his actual experiences, but framed them in a hopeless, helpless narrative from which there was no escape. The severe stress of his combat experience was distilled into a vivid vignette that intruded regularly into his dream world. It was not until he reached his mid-seventies that he finally was diagnosed as suffering from PTSD. The nightmares and waking outbursts of temper that plagued him for so much of his life were finally recognized as symptoms common among soldiers who had experienced combat, particularly those who had to endure captivity as a POW. Here, then, in PTSD, is a perverse consequence of the human capacity for mental time travel. Trauma can be recollected lucidly, in dreams as well as in waking moments, long after nontraumatic, ordinary experiences from early in life are long forgotten.

COGNITIVE APPRAISAL

Another example of a key neocortical intervention in emotion can be found in the attributions made about the source of physiological arousal and the severity of stressors encountered on a daily basis. The interpreter, a system of attention, language, and causal inference, plays an important role in the link between stressors and health. Just as the causes of external events are assessed by the interpreter, the sources of inner emotional arousal are also appraised. If the source of a rapid heart rate and perspiration is attributed by the interpreter to a negative emotion, such as anger or fear, welling up in the body, then the felt emotional intensity is stronger than if some external cause for the arousal is blamed.[16] For example, after just finishing a two-hundred-yard sprint, the heart races and the body perspires, but the arousal is regarded by the interpreter as benign and nonemotional. The appraisals carried out by the interpreter can even color the valence of an emotional experience. If the interpreter assigns the cause of trembling to a threat, then fear ensues. The same trembling, after receiving positive news, would be interpreted as the bodily component of overwhelming joy. Similarly, tears can flow in happiness and in sadness, depending on the causal appraisal carried out by the interpreter.

Even the intensity of the primal emotion of fear can be reduced if the physiological arousal the person is experiencing can be attributed to some benign external source. For example, one study induced fear in college students by leading them to believe that they were about to receive an electric shock. The results showed that "subjects anticipating electric shock spent less time attempting to avoid the shock if they were led to believe that their arousal was due to white noise."[17] The white noise became for them a plausible explanation for why they were feeling uncomfortable, and this interpretation of their feelings led to reduction in fear.

Stressors are also appraised by the interpreter for their intensity and potential for damage. Are there financial resources that can reduce the jeopardy of being hit by an earthquake or a tornado? Are there friends, family members, or church members who can provide comfort at the death of a loved one? Is there any way to control the intensity of the stressor to reduce its effects or is one helpless in the face of it? Is it expected that the stressor is only temporary or is it instead likely to afflict one for weeks, months, or even years? The consequences of exposure to stressors can vary widely depending on the appraisals made. The self-talk generated during the course of cognitive appraisals has enormous consequences for how well people cope with the stressors of life.

As noted earlier in the book, the interpreter typically tries to talk about the world in the best possible light. These optimistic interpretations are called positive illusions. Strong positive illusions about health are related to happiness, contentment, and the capacity to function under duress.[18] A strong belief in the capacity for self-control allows one to take steps to reduce the number and intensity of stressors in one's life. Without a belief that one can effect a change in the environment, it is difficult to mount an effective response. Optimism about the future similarly serves the function of shaping our appraisals of current stressors. A strong belief that things will be better in the future makes the present more bearable. Thus, positive illusions regarding mental health preclude a state of helplessness where nothing one can do matters, where circumstances are hopelessly beyond personal control, and where the future looks even worse than the present. When the interpreter functions in typical fashion by making overly optimistic and rosy attributions about events and the self, it is certainly not functioning in a way that is rational and in touch with reality. Yet these positive illusions are adaptive and protective.

A key executive function of working memory is inhibition. In depression, a negative mood can trigger a cycle of increasingly bleaker thoughts that are not easily inhibited. The inability to control the content of working memory plays a pivotal role in the downward spiral of clinical depression.[19] Those at risk for depression manifest a weakness in the executive function of inhibiting irrelevant information, and they are easily distracted by thoughts that would best be ignored or quickly forgotten. The inner dialogue of verbal working memory instead recycles the negative content in a self-destructive pattern of rumination.

Thus, the ultimate effect of a negative emotion such as sadness is directly influenced by the functioning of the prefrontal cortex and working memory.

Those suffering from depression and anxiety disorders can often be treated through a psychotherapy that focuses on changing the negative attributions and ruminations of the interpreter. These are called cognitive behavioral therapies.[20] For example, such therapies attempt to retrain the attributions made about the causes of life's events. Therapists challenge the negative self-talk in an effort to redirect the narrative of the inner voice. The goal of these methods is to instill a more accurate and less debilitating appraisal process and thereby alleviate the individual's anxiety or depression.

In addition to the self-talk and causal attributions of the interpreter, there is evidence that the executive functions of working memory intervene in altering emotional responses. The activation level of the amygdala can be reduced by reappraising a situation that normally would invoke a significant emotional response.[21] For instance, a scene of women crying outside a church could elicit empathic feelings of sadness as one shares in their grief. However, if one is instead instructed to interpret the women as crying out of joy at the end of a wedding ceremony, then this shared grief can turn to shared joy. The lateral prefrontal cortex associated with executive attention has been shown to be activated during these efforts to change an emotion through reinterpretation. The degree of activation in the left lateral prefrontal cortex increased for those who were most successful at reappraising the scene and changing their emotional experience. As Elizabeth Phelps observed, "Reappraisal is similar to viewing the cup as half full as opposed to half empty," and it can "alter the experience of emotion."[22]

Another way is to reduce, if not eliminate, self-talk altogether. The relaxation response is a reduction of the arousal associated with emotion. It can be induced through meditation methods that concentrate executive attention away from distracting thoughts and replace the usual worrisome chatter of the interpreter with a mental focus on a single word or phrase. For example, in a type of Kundalini meditation in yoga practice, meditators "passively observed their breathing and silently repeated the phrase 'sat nam' during inhalations and 'wahe guru' during exhalations."[23] The goal is to focus attention on a single bodily process—breathing in and out—rather than allowing the contents of working memory to wander aimlessly. At the same time, the attributions, explanations, and narratives of the inner voice are at least partly silenced by the repetition of the chosen phrase. Brain alterations seen in functional magnetic resonance images taken during the relaxation response included changes in the amygdala, reflecting control of emotional arousal, and changes in the prefrontal cortex associated with the executive attention network.[24]

To produce the relaxation response, a variety of such phrases can be used, including simply repeating the word "one" or "relax."[25] If the mind wanders from focusing on breathing and repeating the phrase, the meditator simply returns attention and starts over again. It is important not to worry about the mind wandering, because that will only invite the interpreter to assign blame and generate negative self-talk. For example, after realizing that the focus from breathing and word repetition has been lost, the meditator might say, "Oh well" and return to repeating the word or phrase.

The findings on meditation indicate that human working memory and the inner voice of the interpreter are interlocked with emotional arousal. By practicing focused attention on the process of breathing—an elemental, life-sustaining bodily process that typically unfolds outside of awareness—the contents of working memory are directly altered. By thinking about breathing, the usual wandering of thoughts that pop in and out of working memory is diminished. As for verbal working memory, the inner dialogue of the interpreter is replaced with a simple word or phrase. The usual commentary of the interpreter is in part silenced by redirection. Meditation can take a variety of forms beyond those discussed here, but in general it can be seen "as a family of complex emotional attentional regulatory strategies developed for various ends, including the cultivation of well-being and emotional balance."[26] As such, it elegantly demonstrates how the advanced working memory and the interpreter of the human mind are intertwined with emotion.

The most astonishing example of how the interpreter can affect health through positive illusions is the placebo effect.[27] If the brain's interpreter infers that a pill, a potion, a prayer, a ritual dance, or any other medical procedure—whether from the canon of modern medicine, alternative medicine, or shamanistic medicine—is curative, then it can actually work, just because the person holds and activates a strong belief in its power. The placebo effect, then, is a positive illusion generated by a belief and an interpretation. Remarkably, these mental events by themselves can activate events in the brain and body, including the immune system and the pain perception system. Sham pain pills, sham acupuncture, and even sham surgery supposedly undertaken to correct a source of chronic pain have been shown to produce effective analgesia, based on the patient's interpreting the treatment as real and believing in its effectiveness.

A placebo was once conceived as an inert substance with no specific effect. Thus, any benefit from the placebo must come solely from the patient's belief that it can cause a reduction in pain or a reversal of some disease process. Today, medical researchers think of a placebo as a simulation of an active therapy, a way of invoking the benefits of a real medication through a ritual that engages the patient's hopes and expectations. Shelley E. Taylor, in *Positive Illusions*, observed that "the placebo effect is most likely to take place when it is introduced into a carefully crafted theatrical production designed to enhance unrealistic optimism."[28] The doctor or other health practitioner who expresses confidence in the effectiveness of the treatment can slip similar thoughts into the patient's self-talk. To the extent that the left-brain interpreter assigns a causal role to the value of the treatment and begins to narrate its benefits, it can in fact induce the brain to respond positively. Placebos "clearly achieve their greatest success in the absence of real physical disorders," a fairly common circumstance, in that physicians have estimated "that about 65 percent of the patients who seek treatment actually have problems that are primarily emotional in origin."[29]

The power of the placebo effects is well understood by pharmaceutical researchers investigating new medications. The benchmark for establishing effectiveness is comparison of the new medication to a placebo. In other words, some patients receiving placebos get better, too. Only active medications that perform better than a placebo are worth pursuing, particularly because active medications generally carry some risk of side effects whereas placeboes do not. The individual receiving the placebo must expect it to be an active drug and believe that it will work effectively. Thus, in research testing, the effects of new medications are carried out under blind conditions. For ethical reasons, participants in the research are aware that some recipients will get active medication and others will not. But the recipients are blind to which condition they have been assigned. Moreover, the best research design involves double-blind conditions, in which both the person administering the treatment and the recipient are unaware of its true content. By taking this precaution, any expectations about the possible benefits of the treatment are not communicated in subtle ways, through spoken words or through nonverbal body language and gestures.

Looked at from a different perspective, it is understood that a portion of the benefit from a medical intervention comes from the placebo response, the belief that the treatment will help. For example, the use of morphine and other opioid drugs as pain killers is most effective in managing postoperative pain when the patient receives the medication in an open manner.[30] That is to say, the patient knows when the medication is given and knows it is morphine for the purpose of relieving pain. The clinician injects the medication in full view of the patient and verbally suggests that it will help to reduce pain. These contextual elements activate the patient's belief that morphine will quickly bring on a reduction in pain. By contrast, when the same dose of morphine is given under hidden conditions—administered by a machine without the patient's awareness—it is less effective in reducing pain. Similarly, smaller doses of the analgesic medications can achieve pain management more successfully when given openly than when hidden from the patient. The ritual of administering a pain palliative draws the attention of the interpretive system, initiating a placebo response on top of the purely pharmacological effects of the drug on the pain-perception system. What is more, the open administration can further strengthen the patient's belief regarding the effectiveness of morphine as an analgesic. Once the pain begins to fade after receiving the drug, the interpreter would seek an explanation. With open administration it is easy to assign the cause of pain relief to the morphine injection. With hidden administration, the interpreter is left to wonder exactly what caused the reduction in pain; the inference that morphine must have been injected and must have caused the reduction in pain is less certain.

Not everyone responds to placeboes, and the magnitude of response can vary depending on the exact circumstances of its administration. This variability is not surprising when the response is dependent on the interpreter inferring that the treatment is active rather than inert. If an individual is skeptical about the curative or analgesic properties of a treatment, then his or her interpreter is likely neither to attribute any health improvement to the treatment nor to sustain an inner narrative favoring its continued use.

ADDICTION IN THE MODERN MIND

As another example of how parts of the modern mental ensemble alter our emotional experience, consider addiction. The midbrain and limbic forebrain constitute a powerful system of reward that mediates our feelings of pleasure. Rewards are things we seek out because they make us feel good. Sexual orgasm, good-tasting food, and mind-altering drugs such as alcohol and marijuana are some very different examples of rewards that people pursue for the pleasures they provide. Behaviors that are followed by rewards will increase in their frequency—the law of positive reinforcement. As a consequence, behaviors that lead to rewards can readily become habitual. Food and sex are pleasurable because both are needed for human survival and the propagation of the species; yet the same reward system in the brain can also drive substance abuse.

The reward circuit of the mammalian brain, including the human brain, begins in the midbrain, in a region known as the ventral tegmental area, with a neurotransmitter called dopamine.[31] This area is the source of the dopamine found in the nucleus accumbens, deep in the forebrain, as well as in the amygdala and hippocampus of the limbic system, and also in the prefrontal cortex. Another supplier of dopamine is the substantia nigra, which furnishes a separate region near the nucleus accumbens called the dorsal striatum. Both natural rewards, like food and sex, and addictive drugs control the frequency of behaviors by increasing the concentration of dopamine in the nucleus accumbens. The dorsal striatum is involved in the reward circuit, too, as a means of consolidating into long-term memory the specific behavioral sequences that led to a reward in the past.

Dopamine is the common denominator of drugs that are abused for their rewards of pleasure.[32] However, the opiates (opium, morphine, and heroin), cocaine and amphetamines, nicotine, marijuana, and alcohol each have their own unique signatures composed of other neurotransmitters in addition to dopamine. Addiction means that the user becomes dependent on the drug. This is contrasted with the overuse or abuse of drugs. Occasional drug use can lead to abuse and possibly even addiction. The key difference is that abuse does not involve dependence on the drug, although abuse can progress to addiction. To illustrate dependence, consider heroin, a highly addictive opiate. Intravenous injection or smoking of the drug causes intense intoxication. The powerful reward of feeling good increases the likelihood of taking the drug again, as the law of positive reinforcement goes to work on the drug taker. Abstinence from the drug produces strong negative feelings of anxiety, restlessness, and dissatisfaction as well as physical discomfort—these are symptoms of withdrawal from the presence of the drug in the brain. During withdrawal, the drug user craves the drug and becomes preoccupied with obtaining it. The negative emotions of withdrawal are associated with a strong increase in cravings for the drug. Over time, the law of negative reinforcement comes into play. In negative reinforcement, a behavior is increased, in this case drug ingestion, to terminate an aversive stimulus, in this case the anxiety and physical discomfort of withdrawal. Regrettably, negative reinforcement teaches the user to take the drug so as to avoid the aversive state of withdrawal. Although addicts start out taking heroin to obtain its powerful positive reward, they end up taking it simply to avoid the pain of withdrawal. The intoxication is no longer as positively rewarding, but avoidance of withdrawal keeps the addict preoccupied with maintaining a constant supply of heroin. In short, the addict has moved from an impulsive stage of drug use driven by positive reinforcement to a compulsive stage driven by negative reinforcement.

Like humans, laboratory rats can become addicted to heroin, cocaine, or alcohol. This is because humans and rats share in common the reward system of the mammalian nervous system. The manner in which positive reinforcement initially drives the impulsive use of drugs to be later replaced by compulsive use maintained by negative reinforcement can readily be studied in the rat. Even so, there are ways in which human addiction is distinctive. A key executive function of human working memory is the capacity to plan and solve novel problems. This capacity of our advanced working memory system enormously complicates the picture of human addiction. Unless drugs are readily available—like they are for a rat in a laboratory cage—human beings do not simply prepare an injection or take a drink whenever they feel like it.

Instead, they often must devote considerable time and effort to finding and purchasing the drug in the first place. Planning to obtain the drug, and overcoming any obstacles to do so, is part and parcel of human addictions, which simply have no parallel in nonhuman species.

> Human addicts face a situation different from rats that merely lever-press for drugs. . . . An addict who steals, another who scams, another who has the money and simply must negotiate a drug purchase—all face new and unique challenges with each new victim and negotiation. Instrumental ingenuity and variation are central to addictive drug pursuit in real life.[33]

The powerful working memory system of human beings also complicates the experience of cravings for illicit drugs, alcohol, cigarettes, and other addictive substances. As already discussed, craving—the incessant motivation to obtain and ingest the substance—is particularly severe during the compulsive stage of drug addiction. It is also prevalent in compulsive gambling, compulsive sexual behavior, and compulsive eating. For instance, "food cravings have been shown to trigger binge-eating episodes, which in turn contribute to both obesity and disordered eating, especially bulimia nervosa—increasingly serious problems for Western societies."[34]

The visual and spatial stores of human working memory allow us to maintain transient mental images for a period of time. By allocating executive attention to these mental representations, it is possible to keep them active and constantly in mind. Imagery of desired foods, drinks, or smokes become very difficult to get out of one's mind as the cravings for them occupy more and more of the resources of working memory. In one study, people were asked to imagine that they were eating their favorite food as a way to induce cravings. The results showed that "the strength of participants' cravings was related to the vividness with which they imagined this scenario."[35] In other words, an effective use of visual-spatial working memory led to stronger feelings of cravings. Another study showed that cravings consumed attention, too, with participants not being able to perform a concurrent task as well when they were experiencing cravings as when they were not. Ironically, it is our human knack for keeping things in mind that works against us in compulsive behaviors. Cravings would dissipate quickly if images could not be so readily maintained in human working memory.

JEALOUSY

In 1899, in the city of St. Louis, a young woman, Francis Baker, murdered her boyfriend, who was apparently cheating with another woman. The incident is immortalized in the popular song "Frankie and Johnny" written in the early twentieth century, and in a mural painted by Thomas Hart Benton, which still hangs in the House Lounge of the Missouri State Capitol in Jefferson City. More than one hundred years later, Clara Harris became known as the "Mercedes murderer" for running over her husband multiple times with a car after discovering him at a hotel with a mistress. In the *American Scientist*, Christine Harris recounted these historic murders as illustrative of the intensity of jealousy, noting that sexual betrayal is a relatively common reason for intentional homicide, usually turning up in the top three motives.[36]

Men more commonly murder their wives or girlfriends than the reverse, but this does not seem to be due to men being more likely than women to be incited to murder by jealousy. Rather, it stems from the simple fact that males commit more violent crimes of all types. In a comparison of homicides committed by men and women in six different cities in the United States plus five other countries (Africa, Canada, India, Poland, and Scotland), jealousy turned out to be as likely a motive for women as for men.[37]

Jealousy is a powerful emotional signal from the limbic system. Seeing one's love interest interact with a potential romantic rival can trigger an aversive emotional state. When the threat posed by a rival is seen as real and serious, the intensity of jealous feelings—a strange blend of fear and anger and betrayal and sadness, all rolled into one monstrous mess—can reach extremes. In the morbidly jealous, those feelings come with a full-blown delusion that their mate is cheating despite little, if any, objective evidence of infidelity. Morbid jealousy is "frequently accompanied by anger, depression, and urges to check up on and spy on . . . mates."[38] Even in cases where jealousy has not reached the level of delusion and clinical abnormality, it can nonetheless prompt rage leading to assaults and even murder.

Jealousy highlights the importance of the interpreter in human emotion. A male directing his gaze steadily at another attractive woman can arouse some degree of jealousy in his female partner. A flirtatious conversation between his wife or girlfriend and another good-looking man can do the same in a male. The explanations generated for these observed events, and the inner dialogue that ensues, determines the degree to which jealousy is aroused. Christine Harris has argued that

> the primary appraisal of threat might be elicited by an input as simple as a positive interaction between the beloved and any potential rival (in sitcom terms, an act of gallantry or a sideways glance at a swinging skirt). Such an interaction between two others may elicit a vague sense of threat that does not have to be consciously assessed, may be innate and may occur in other animals. It functions to motivate actions that will break up the threatening liaison. At least in human adults, additional appraisals also come into play, including efforts to figure what the liaison implies for one's relationship and oneself. These appraisals affect both the intensity and direction of jealous feelings.[39]

Even in cases of normal jealousy, and certainly in cases of morbid jealousy, the interpreter is occupied by drawing inferences from minimal cues. The potential threats are unlimited in number and severity. Eye contact, smiling, and hugging are all normal forms of human social interaction that do not necessarily imply a threat of emotional or sexual betrayal. Yet they are also ways of flirting that may lead with time to infidelity. How these signs are interpreted determines whether jealousy is aroused as well as its degree of intensity.

Mental time travel also develops and shapes jealousy. Reconstruction of a past incident of flirtation can become distorted and blown out of proportion. As with other kinds of episodic memories, recollection is not necessarily veridical. If one is currently feeling jealous, then it can trigger reconstructions of past episodes that augment the feelings of anger and betrayal. Imagining future episodes of flirtation and infidelity comes equally easy because of our capacity for mental time travel. Fictitious rivals can be conjured in the imagination just as easily as real ones. Indeed, for the morbidly jealous individual, the imaginary rivals constitute much of the evidence for the delusional conviction that his or her mate is unfaithful.

Chapter 8

THE SOCIAL MIND

The Great Wall of China was completed in the third century BCE as a defense against the nomadic warriors to the region's north. At two thousand miles in length, it symbolizes the "us versus them" mentality ingrained in the social circuitry of the human brain. We have a strong need to erect barriers between us and groups that we perceive as different, whether those differences are defined by the shade of skin color, the texture of hair, the shape of facial features, the clothing worn, the food eaten, or even the utensils by which food is eaten. Ethnicity and culture have always divided us as a species, and the social mind cannot be plumbed without diving straight into this history of divisiveness.

Yet, at the same time, it is equally true that human beings cannot live in isolation from one another. John Donne immortalized our dependence on each other, writing, "No man is an island, entire of itself; every man is a piece of the continent, a part of the main."[1] Indeed, our self-concept—the central organizing structure of semantic memory—is defined in large measure by how the self fits into not one but many groups of other human beings. Social identification starts in the immediacy of one's nuclear family and close friends, but it extends further into far larger groups. Imagine a set of concentric circles radiating outward, encompassing larger and larger sets of people. A person is a member of a family, an extended family, perhaps several organizations in one's local community (e.g., a neighborhood, a church, a school, a service organization), a town, a city or greater metropolitan area, a state, and a country. Ultimately, we are all citizens of planet earth, but the immense size of this group—more than seven billion—seems too large and abstract for the human mind to readily grasp. The ties that bind us as fellow voyagers on a planet around the sun, all with the same utter dependency on the earth's capability to sustain human life, are regrettably fragile. When push comes to shove, people regard the needs of their country, their city, or their family as more pressing than the needs of humanity as a whole. The need to belong with others is part of the fundamental design of the human mind. "I'm with them" is stated loudly and clearly in what we say, what we wear, and how we behave.

Thus, the neural circuitry of affiliation with others and need for unity is as much a part of the social mind as the divisions and prejudice that can separate "us" from "the other." In a way, they are two sides of the same coin. People stress their similarities to other members of their group while at the same time noting their differences with other groups. In so doing, they sharpen the category boundary between "us" and "them." As would be expected by the out-of-Africa hypothesis of a common origin for modern human beings, there is considerable overlap in the gene pools of human populations around the globe. By far, the greatest degree of variability in the human genome lies within populations that are thought about as constituting single ethnic or racial groups. What follows is a thought experiment that illustrates the point that most variations among all individuals in the world can be found within any single population:

> Suppose a grand catastrophe occurred (e.g., a large asteroid struck Earth) that resulted in the extinction of all of humanity except for a population of Eskimos living in the Arctic Circle. After many years, this population expands and migrates to eventually resettle the entire planet. From such limited origins, humanity would now have fully 85% to 95% of its pre-cataclysmic genetic diversity. People on different corners of the planet may look more similar to one another than they did before the calamity, but the genes for visible morphology are only a very small—and inconsequential—part of the totality of human diversity.[2]

Although the features of human physiognomy are trivial in the context of the genome, they are readily perceived and highly accessible to thought. So, it is hardly surprising that racial groups are deeply grounded in our psychology, if not our biology. The human social mind categorizes, ranks, and judges other human beings on the basis of surface appearance. Images of how a person looks can be maintained in the visual-spatial stores of working memory and reflected upon in relation to language and behavior. Ethnic differences in language, religion, politics, dressing, housing, and eating divide groups of human beings in tandem with the visible features of the face and body. By contrast, the shared genomic commonality of different ethnic populations is not so easily perceived, thought about, and remembered. It is known to us only as an abstract concept from a complex scientific discipline. And, even then, this knowledge has been disseminated only briefly, since the time that the human genome was first deciphered and comparisons were made of different world populations.

GROUPS

People, regardless of their geographic location and culture, belong to groups of other human beings. Although the world population is now staggering in size, measured in the billions, the day to day interactions of human beings are confined to relatively small groups of people. Human beings—with the rare exceptions of hermits—seek out associations with others in their work and in their leisure. Face-to-face interactions, joint activities, and conversations take place on a periodically recurring, if not daily, basis. With social networks now available via the Internet, these linkages with others are less geographically constrained.

The psychological importance of the bonds we share with others can be seen when they must be broken. Breaking up an established relationship is highly distressing to human beings, and this fact is again universal across different cultures and different age groups. Although this is most obviously true of romantic relationships, it is also true of other friendships. For example, consider the resistance that friends have in separating, such as when graduating from school or moving to a different city:

> As such transitions approach, people commonly get together formally and informally and promise to remain in contact, to share meals or other social occasions together, to write and call each other, and to continue the relation-

ship in other ways. They also cry or show other signs of distress over the impending separation. . . . Reunions constitute an occasion for people to see former acquaintances. The massive exchange of greeting cards during the Christmas holiday season includes many cases in which the card is the sole contact two people have had in the entire year, but people still resist dropping each other's name from the mailing list because to do so signifies a final dissolution of the social bond.[3]

Although we resist breaking social bonds with others, there are times when we are excluded regardless of our desire to continue the relationship. Being dumped by one's partner in a romantic relationship can be an intensely agonizing experience. The depression brought on by such rejection is convincing evidence of the centrality of our need for belonging and acceptance. One might wonder, however, if the person who rejects the love of another is somehow above this fundamental need. The answer seems to be negative. Typically, romantic relationships dissolve because the rejecter finds his or her partner unattractive or incompatible in some manner and unable to satisfy the desire for a close relationship. Or the rejecter has already satisfied this need with another preferred partner. Even in these cases, the rejecter often experiences feelings of guilt and empathy at the pain he or she has caused by breaking off the relationship.[4]

Cognitive neuroscientists have investigated the brain circuits that underlie social rejection. Intriguingly, the distress felt when excluded from others triggers the same system that mediates the perception of physical pain.[5] The anterior cingulate cortex (ACC) is part of the neural circuitry for the executive attention component of working memory. A key function of this neural region is to monitor for conflicts in cognition and emotion. For example, when two cognitive representations are simultaneously active but call for actions that are diametrically opposed (e.g., go left and go right), the ACC detects this conflict. It acts, in other words, as an alarm system that informs the rest of the brain that something is wrong and needs attention. The dorsal portion of the ACC is known to signal perhaps the most primitive kind of alarm in the form of physical pain felt in the body. Pain is the brain's way of making sure one is aware that something is wrong with the body and that our attention is required. Just as physical pain is important for our survival, could it be that the dorsal ACC also is activated when social attachments are disrupted. To the extent that our need to belong with others is fundamental to our security and survival, it makes sense that the social-attachment system might co-opt the existing physical-pain system.

To study social exclusion in the lab, psychologists had college students play a simple game of catch on a computer.[6] The game is called CyberBall and the participant's avatar plays catch with the avatars of supposedly two other people in the game. In actuality, the other two avatars are controlled by the computer rather than by people. The game starts out well, with the ball being tossed back and forth amiably among all three participants for seven throws. But then the real participant is excluded from the game for the remaining forty-five throws. The other two avatars hog the ball and never toss it back to the real participant. The fMRI scans taken during this exclusion period revealed a high degree of activation of the dorsal ACC. Another region was also found especially active during social exclusion, namely, the right ventral prefrontal cortex. This is of interest because this same region seems to mediate reductions in the perception of pain caused through a placebo effect, when people gain relief simply because they believe a pill is a pain killer. Presumably this prefrontal region is brought into play to inhibit the distress felt by social exclusion and to dampen the warning signals emitting from the ACC.

Family and friends in our closest social groups provide a means for coping with the everyday ups and downs of life, as well as with the more severe major life events of job loss, family deaths, and accidents. Fight or flight is the well-known response to stressors in life. Far less well-known is our need to belong and draw strength from close social relationships. Along with fight or flight, human beings also tend and befriend. A friend can be trusted with one's deepest thoughts and feelings, and can be a source of support on par with the first attachment of mother to child. Shelley E. Taylor has argued that "under conditions of stress, tending to offspring and affiliating with others ('befriending') are at least as common responses to stress in humans as fight-or-flight."[7]

That human beings turn to one another during times of difficulty provides a compelling illustration of how important social connectedness is for our species. When positive social relationships are ruptured the brain triggers both a neurohormonal response—elevation of oxytocin—and a social-behavioral response—elevation of seeking affiliations.[8] Oxytocin, as well as the general reward system involving dopamine and opioids, serves to reinforce successful social contacts with others. Stress is directly reduced by tending and befriending, eliminating the problem that triggered the chain of events in the first place. On the other hand, what if others reject one's attempts to affiliate? Such negative social contacts will only worsen the stress, with negative consequences for health and well-being.

The need to belong would appear to be a genetically encoded result of our evolutionary history. It makes sense that human beings could best survive and reproduce in groups. There would be several advantages:

> Groups can share food, provide mates, and help care for offspring (including orphans). Some survival tasks, such as hunting large animals or maintaining defensive vigilance against predatory enemies, are best accomplished by group cooperation. Children who desired to stay together with adults (and who would resist being left alone) would be more likely to survive until their reproductive years than other children because they would be more likely to receive care and food as well as protection.[9]

Human beings by nature seek to belong to social groups and to form long-term relationships with others. People are motivated to maintain social relationships and try to avoid needlessly breaking them off. When under stress, tending to these relationships provides a means for coping. Finally, rejection by others even invokes pain in regions of the brain that also mediate the perception of physical pain.

The human need to belong was very likely shaped and constrained by our advanced working memory ability and by language. In the social domain, working memory is needed to keep track of all the individuals with whom one has a relationship. The number of these relationships that can be successfully monitored and maintained would depend on the capacity of working memory and in particular on executive attention. It turns out that a form of the encephalization quotient that takes into account the size of the brain relative to body size is a good predictor of the typical group size favored by various species. That is to say, the volume of the neocortex relative to the volume of the brain stem predicts the size of the group that can be effectively monitored.[10] Because of our neocortex, especially with respect to the prefrontal cortex mediating the executive functions of working memory, human beings can monitor relationships in fairly large groups. Based on our neocortex ratio, a group size of nearly 150 is expected, and "there is considerable evidence that groupings of this size occur frequently in modern and historical human societies."[11] This is a relatively large group size compared with other primates with much smaller neocortex ratios. These nonhuman primates maintain social relationships through grooming each other. By contrast, human beings can use language to achieve the same purpose. By sharing information through language, interpersonal relationships and the cohesion of the group can be preserved. Thus, our unique ability to use language possibly allowed for larger group sizes than could reasonably be maintained through social grooming. Conceivably, "a form of social interaction that was more efficient than grooming in its use of time would thus have been required to facilitate the cohesion of such large groups."[12]

US VERSUS THEM

In his 1954 book *The Nature of Prejudice*, Gordon Allport pointed out the natural inclination of human beings to divide themselves into homogeneous groups, into "us" versus "them" categories. He regarded this social categorization as a normal extension of our ability to form and use categories about everything. Just as we group together chairs and see them as different from tables, the same categorization process is applied to groups of human beings. Allport's insights have been confirmed and extended by social psychologists over the past half century.[13]

Categorization is a fundamental aspect of cognition. Both human and nonhuman minds organize objects, events, and life forms into groups based on their perceptual similarity. Categories simplify our ability to cope with a complex world by allowing us to ignore differences that do not make a difference. Instead of treating every object, event, or plant or animal that we encounter as a unique individual, it is easier to treat each as a member of a category. In this way we can quickly and automatically perceive, understand,

and respond to the things we encounter. Categories thus guide our interactions with the environment with a minimum of cognitive effort. So it should not surprise us that we categorize other members of our species into easily identifiable groups on the basis of perceptual characteristics. Skin color, facial features, and hair texture provide means for lumping together individuals into "racial" categories, just as sex characteristics divide us into male and female gender categories.

Knowing who belongs to one's group—who is familiar—is a basic cognitive function adapted for survival. The unfamiliar is inherently risky even to the point of being potentially life threatening. The amygdala buried in the limbic system serves as an early warning device for novelty, precisely so that attention can be mobilized to alert the mind to a potential danger and to prepare for a potential response of flight or fight. Categorization of other human beings as either familiar members of the in-group or novel members of the out-group provided a simple cue for triggering the amygdala. Neuroimaging of the amygdala reveals greater activity when faces of the out-group are presented, even when the racial identity was irrelevant to the task.[14] For example, when asked to judge the gender of a series of faces, Caucasian Americans revealed heightened amygdala activations when viewing African American faces. This was especially true for White individuals whose implicit or unconscious racial attitudes were biased against Blacks, though it is interesting to note that their explicit or conscious attitudes were unrelated.

The same finding occurs when Whites categorize Black faces according to their age. Whether it is age or gender, both judgments aim to put someone quickly into a social category. As Allport observed, this kind of quick sizing up of another person is useful for getting a fast assessment of how to interact with the person—whether to approach or avoid at the most basic level. But it is also possible to perceive the faces of other human beings with the goal of seeing in what way the person is unique as an individual rather than as a member of a category. When the study was repeated with a goal of individuation, the amygdala was not activated at an elevated level when Whites perceived Black faces.[15] To encourage individuation, the participants in the study were asked to decide whether the person viewed would like or dislike a particular vegetable, such as asparagus, broccoli, carrot, and so on. This judgment required one to think about individual personality conveyed by the facial features. It was not a simple categorization of gender or age, but a complex elaboration of a person's likes and dislikes. The goal held in working memory thus inhibited an otherwise automatic activation of the amygdala.

The capacity of modern human beings to use symbols, including words, in their thinking facilitates divisions into us versus them. Flags as symbols of nationalism are an illustration. When the American flag, for example, is desecrated by members of an out-group, it is regarded as an attack on the nation by those with strong feelings of loyalty and patriotism. It is not just a piece of cloth. An attack on the flag is a symbolic attack on the in-group. Logos, uniforms, emblems, or pins worn on clothing serve similar symbolic functions of strengthening feelings for members of one group. This is why we have school colors, for example. Like the arbitrary nature of words in a language, the symbols of group identification are similarly arbitrary or at least accidents of history. The Democratic Donkey and the Republican Elephant certainly have a history, but it is hard to argue that they are logical choices. Similarly, it appears arbitrary to associate blue states with Democratic control and red states with Republican control.

Symbols alone can be used to foster a feeling of group identification and loyalty. Social psychologists can study the us-versus-them phenomenon in a controlled laboratory experiment by forming novel groups that have nothing to do with real-life group identity. Members are randomly selected and then given a common symbolic identity, such as a name or a symbol to wear, and the symbol drives the two groups apart. This effect comes so naturally to human beings that it can readily be seen even in children. For example, in one study grade-school children were randomly assigned to one of two groups that differed only in tee-shirt color.[16] Half of the students were selected to wear a red shirt and half a blue shirt. At the outset, students in the red group were no different in their attitudes or views of each other, because they were after all randomly assigned to one group or the other. However, over time the symbol of shirt color was perceptually salient and became a basis for group identification as the teachers used the color groups to organize the classes in terms of their activities, bulletin boards, and the like. After several weeks of the experiment, the "Reds" came to see themselves as more similar to one another and different from the "Blues," who in turn showed the same in-group bias. Each group evaluated its own competence and performance as superior to the other group.

In Nazi Germany, the out-group Poles and Jews were demonized through language. Through a deliberate program of propaganda, the state described Jews as less than human: *Untermenschen*. Using the symbols of graphic posters, Jews were identified as vermin who deserved the scorn of society as much as rats in the sewer. Yet the Nazi definition of "the other" as a biological category was doomed to failure, as even science of the 1930s had begun to hint. Although DNA had not yet been discovered and genome comparison was unknown, German scientists knew they had a problem on their hands. To separate Aryans from Jews required the explicit use of symbols because the biological phenotypic distinctions were so shaky. As Mark Mazower described in *Hitler's Empire:*

> The "Breslau school" believed in tracking blue eyes and blond hair, but Otto Reche and Fritz Lenz—two luminaries of academic racism—thought physical characteristics were crude markers since most individuals were themselves mixed racially. For Hans Günther, a popularizer of Nazi science, even Germany contained strains of all the major European races—the Nordic, East Baltic, Alpine and Dinaric—as well as fortunately small quantities of Mediterranean and Inner Asian blood. . . . A few heretics solved the problem by matching up the categories of race and *Volk* by talking about a "German race," but this was criticized by most academics as unscientific. . . . All of this spelled enormous confusion, regarding not only the Germans but also the Jews.[17]

It was difficult to tell Germans from Poles from appearance, so difficult in fact that the Nazi regime forced Poles to wear markers, such as a violet letter *P*, so that the Germans could avoid fraternizing with them. Similarly, those of Jewish descent were required to wear a yellow Star of David as a means of identification. A truly perverse result of our symbolic capacities is our rendering them in the service of racist ideology. Without the help of such symbols, the pogrom would have been hobbled for lack of convenient ways of identifying "the other."

Language provides a simple dichotomy for keeping straight our in-group bias. We label the in-group as good and the out-group as bad. The Implicit

Association Test (IAT) capitalizes on this labeling process as a way to identify the degree to which people are biased toward or against various groups. For example, in the IAT for race, people make rapid judgments about pictures of faces and words. Half are White faces and half are Black. Half of the words are words associated with positive meanings (e.g., joy, love, peace, happy) while the other half are words associated with unpleasant meanings (e.g., agony, terrible, horrible, evil). The test measures how quickly one can process the faces and words when Black faces are paired with good versus bad compared with when White faces are similarly paired. The idea is to detect whether there is a natural association that speeds processing when one race or the other is paired with good versus bad. This test can be used to detect racial bias against either Blacks or Whites. For example, bias against African Americans would be detected if Black (but not White) faces are responded to more slowly when they are paired with good words compared with bad words. The person taking the test is typically unaware that his or her responses are fifty, one hundred, or two hundred milliseconds slower when Blacks are linked to good compared to when they are linked to bad. Test takers may not even consciously report any racial prejudice at all. Compared to IAT measures, self-reported measures of racial attitudes turn out to be poorer predictors of actual interracial behavior.[18] This is not generally the case, for biases in attitudes and behavior regarding consumer products or political preferences are readily predictable by just asking people what they think.

The language used to derogate other human beings serves to reify the abstraction of "the other" into a concrete subhuman or demon. Words allow us to hold concepts in verbal working memory that can be reflected upon, judged, and despised. The words themselves amplify the prejudice by becoming an integral part of one's thoughts and interpretations. A member of an out-group can become more than just another disliked human being; hateful language can reify the out-group member such that he or she comes to be seen as truly a monster. Seeing the target of one's prejudice as just another human being is hard when the mental image is shrouded in hateful words that are so easily brought to mind. Even at the conclusion of the most horrific violence the world had yet known, the delegates participating in the 1919 peace settlement at Versailles, which ended World War I, denigrated the peoples of whole regions of the world. The racist language called for colonizing lands in the Middle East, Africa, the subcontinent of Asia, and the Pacific on the grounds that the inhabitants were not yet capable of independence.

> Legal and political theorists talked about tiers of sovereignty and they distinguished between "civilized," "barbarian" and "savage" people. . . . At Versailles the victor powers had bestowed sovereignty upon the "civilized" peoples of eastern Europe and created a set of "New States" there, subjected only to the conditional oversight of the minority rights regime. In the Middle East, they had established League mandates to usher the Arab peoples towards independence and full statehood, a process that brought freedom (of a kind) to Egypt and Iraq before the outbreak of the war (though not to Syria or Lebanon). Only among the "savages" of Africa and the Pacific did they justify colonial rule into the indefinite future.[19]

The virulence of prejudice is amplified into hatred and violence because of the interactive effects of the modern ensemble of mental parts. When racial prejudice drives human beings to find in the out-group a scapegoat for all the problems of a nation, hatred and horrific violence can follow. What begins as apprehension and fear of "the other" can end in dehumanizing and even murdering entire populations of human beings in acts of genocide, as human history keeps reminding us. The intensity of intergroup conflict in human beings is a direct consequence of the very mental capabilities that set us above other species. The planning of revenge for past grievances would not be possible without the executive functions of human working memory. Only human beings are capable of symbolically representing the in-group and the out-group, through flags, coats of arms, and other visual images, and of characterizing the good of the in-group and the evil of the out-group in the abstractions of language. Only human beings remember the atrocities encountered earlier in their lives by mentally traveling backward in time to reenact the pain all over again. Through oral and written history, our unique capacity for language can keep alive hurtful memories possibly indefinitely and certainly for centuries. Only human beings can commemorate the dead and keep alive the hate through the symbols of monuments, paintings, texts, and other external forms of memory. In short, the mental ensemble of the modern mind is the reason for the culture that sets us free from a purely animal existence and, at the same time, ensnares us in conflict and violence far exceeding any found elsewhere in the animal kingdom.

THE FUNDAMENTAL ATTRIBUTION ERROR

The usual causal inferences ground out by the left-hemisphere interpreter are integral to racial prejudice. The interpreter is biased in its own way. It tends to see the causes underlying the self's behavior as changeable and situation specific. By contrast, the interpreter assigns explanations to the behavior of other people based on permanent and pervasive factors, such as the fixed, immutable traits of biology and personality. If you cut off another car in heavy traffic on the freeway, your interpreter explains that it was an accident, a temporary lapse of attention, one never to be repeated. But when another person cuts you off in the same situation, the interpreter infers without hesitation that the other driver is a lunatic, has an antisocial personality, or should have his or her license revoked. Although we attribute to our own behaviors some transient, situational cause, the interpreter sees nothing but unvarying, permanent traits as the cause of other people's behaviors. This *fundamental attribution error*—as social psychologists label it—will prompt one to see members of the out-group as bad people because of their biology and permanent disposition, not because of the circumstances in which they temporarily find themselves.

The fundamental attribution error refers to the human tendency to interpret the cause of other people's behavior in terms of dispositional traits. For example, ability, personality, sex, and race are enduring characteristics of the other person, and these would seem to be a plausible explanation for the person's behavior. It is an attribution error, however, because it fails to take into account how factors external to other people can have a major impact on their behavior. These situational factors are temporary rather than permanent, and consequently it is quite possible that the person's behavior will change in the future if his or her situation changes. This error of interpretation is pervasive, so much so that it is regarded as perhaps the fundamental principle of social psychology. Consider it yourself:

Are you shy or outgoing, or does your behavior depend on the situation? Are you calm or intense, quiet or talkative, lenient or firm? Or, again, does your behavior depend on the situation? Now think of a friend, and answer the same questions about his or her behavior. Do you notice a difference? Chances are you do. Research shows that people are more likely to say "It depends on the situation" to describe themselves than to describe others.[20]

If it is correct that the left-hemisphere interpreter is responsible for the fundamental attribution error, then it ought to be a universal characteristic of human beings, rooted in brain structure since the origin of modern human beings. Cultural differences should not alter the likelihood of making the error. This question has been of interest in social and cultural psychology because it is known that in collectivist cultures the self is defined largely by relationships to other members of one's reference groups. Collectivist cultures see the self as part of a group more so than do individualistic cultures. If one lacked an independent view of the self, then there may be less of a tendency to assume that the behavior of another person corresponds to his or her personality. The dispositional error of downplaying the role of temporary, varying situational explanations for behavior may be less likely for those who belong to a collectivist culture. Indeed, there is some support for this hypothesis: "Perceivers from individualistic cultures (e.g., Australia, Great Britain, United States) tend to favor dispositional explanations for behavior, whereas perceivers in collectivist cultures (e.g., China, India, Taiwan) tend to prefer situational explanations."[21]

However, asking people to compare a dispositional explanation (e.g., he said little at the meeting because he is shy) with a situational explanation (e.g., he said little at the meeting because he had a headache) is one thing; perhaps cultural differences in fact lead one to *prefer* one kind of explanation over another, depending on the stress placed on individualism and the independence of the self from the group. But it may still be the case that all human beings, regardless of their cultural identity, still actually make the dispositional attribution. A comparison of American and Chinese participants found that both individualists and collectivists in fact are equally likely to make the fundamental attribution error.[22] After reading an essay in favor of capital punishment, both ethnic groups assumed that the author's true attitude corresponded to the position advocated in the essay, even though they had been warned that the essayist was assigned a position to defend. In line with the view advocated here, the left-hemisphere interpreter is responsible for the fundamental attribution error, and the fundamental attribution error is truly fundamental to the modern mental ensemble of our species without respect to culture.

A fundamental attribution error with immense historical consequences was the one made by Adolph Hitler. His hatred of the Jewish people was driven by an attribution of biological inferiority. In his imagination, his *Volk*—the in-group—was a superior strain of humanity—the Aryan race. The Aryans, according to Hitler's demagoguery, were destined to rule Europe and the world, at the expense of the non-Aryan Poles and Jews. Hitler was not alone in his belief that people could be and should be ranked. It was a standard belief of Western science, pervasive throughout Europe and the United States. Herbert Spencer's concept of the survival of the fittest and Charles Darwin's concept of natural selection were widely interpreted as implying a ladder of success. Life evolved from low forms to higher and higher forms in a great chain of being. As Steven Jay Gould documented in his book *The Mismeasure of Man*, post-Darwin, "subsequent arguments for slavery, colonialism, racial differences, class structures, and sex roles would go forth under the banner of science."[23] The leading scientists of the day, Francis Galton, Samuel George Morton, and Paul Broca, contributed their measurements of the skull and the brain to demonstrate racial differences.

> The leaders of craniometry were not conscious political ideologues. They regarded themselves as servants of their numbers, apostles of objectivity. And they confirmed all the common prejudices of comfortable white males—that blacks, women, and poor people occupy their subordinate roles by the harsh dictates of nature.[24]

Although it was understood by all that human beings occupied the top of the chain relative to other species, it was equally assumed by the Caucasian scientists of the day that their racial group held the loftiest position of all.

The scientists committed the fundamental attribution error by assuming that other human races were inferior for fixed, unforgiving biological reasons. Non-Caucasians, they argued, were obviously inferior in intellect because their brains were smaller. Although social Darwinism invited and encouraged the idea of ranking categories of human beings, a leading biologist of the era had already asserted the inferiority of non-Caucasian races nearly a decade before Darwin published the *Origin of Species*. In 1850, the famous naturalist and professor of zoology at Harvard University, Louis Agassiz, claimed that the Genesis account of Adam applied only to Caucasians.[25] Thus, the racial science of Nazi Germany that came later certainly did not invent the notion of a superior race. Their innovation was the still more self-centered claim that members of the "Aryan race," conceived as non-Jewish Caucasians, especially those with Nordic blood lines, were destined to dominate history.

The human genome project has now laid bare the nonsense of such claims of racial superiority. The out-of-Africa model supported by contemporary genetics asserts that we are all close genetic relatives, despite some relatively minor differences in physiognomy. Across the globe, 85 to 90 percent of total genetic variation in human beings is found within any single population, such as Asians, Africans, or Caucasians.[26] Individual differences reflecting variation between racial groups amounts to only 10 to 15 percent of the total variation. The relatively small variations attributable to group differences explain the perceptually obvious differences in facial features, such as the shape of the eyes and nose, hair texture, and skin color, as well as less visible traits, such as predispositions toward particular diseases or medical conditions (e.g., sickle-cell anemia or lactose intolerance). But the differences are far too small for biologists to classify all humans as anything but a single species, without meaningful divisions or subspecies. Biologists define subspecies or races in a strict biological sense when 25 to 30 percent of the variation is between groups; indeed, in clear cases of racial subgroups in other mammalian species, the figure reaches 60 to 80 percent.[27] Although cultural or ethnic differences often divide peoples from different regions of the world, our genetic heritage largely fails to do so.

Think of different breeds of dogs as true racial groups in the biological sense. A sheltie is not a boxer, precisely because each racial group of dogs has been carefully bred to achieve a unique genetic profile with a distinctive phenotypic appearance and temperament. The gene pools from populations of *Homo sapiens* located in different geographical regions show far too much overlap to be considered true biological races.

Even so, our ethnic and cultural differences are deeply embedded in the psychology of us versus them. The prejudice against those in the out-group is furthermore aggravated by the dispositional attribution made by the interpreter of the left hemisphere. The profound error of seeing "the other" as biologically distinct from us is a root cause of human misery. History has repeatedly shown that when societies are structured around racial categories and their ranking, trauma, poverty, and despair are in store for human beings in the out-group at the bottom of the ranking.

But it was not just ignorance of the human genome that contributed to the dispositional error made by the nineteenth-century scientists. Their racist rankings of brain size reveal to us how the interpreter of the left hemisphere could twist facts and dismiss contradictory evidence to affirm preconceived beliefs. As Stephen Jay Gould argued, the scientists did not commit fraud so much as self-deception. A hierarchy of the races supposedly emerged from measurements of cranial capacity. The idea was that a larger brain implied more intelligence, with White males topping all. Yet the measurements were suspicious and prone to subjective bias. For example, with measurements made by filling a skull specimen with mustard seed, one might easily overfill a White skull by packing the seed tightly. By contrast, knowing that a large Black skull would not fit one's preconceptions, filling the skull quickly with a few shakes, and most certainly no packing, would result in less volume of seed.[28] In other cases, the data were selectively recorded or dismissed to fit racial preconceptions. By electing to throw out the smaller skull sizes of Hindus from his overall sample of Caucasians, a larger cranial capacity could be claimed for Causasians.[29] This manipulation of the data was ironically justified on the grounds that the Hindu skulls from India were smaller than those of other nations. Working in the opposite direction, an unusually large number of Inca Peruvians were selectively included in the calculations because they tend to have smaller heads in general, and this pulled down the average for American Indians. Thus, by selective sampling it was possible to "prove" that Caucasians have larger cranial capacities than do Indians. Of course, this was achieved by simply ignoring all the counterexamples. Were it possible to capture and replay the inner dialogue of such a scientist at work, doubtless the self-talk of his interpreter provided a convincing explanation of why the selective sampling was the right thing to do.

The males who dominated science in the nineteenth century also concluded that male brains were larger than female brains; it followed that males were considered more intelligent. Again, this patriarchal conclusion was reached by misinterpreting the evidence. Specifically, correction factors were selectively applied.[30] For example, when confronted with an atypically large non-White male specimen, a correction factor of body size was introduced into the reasoning to explain away the so-called anomaly. Namely, it was duly noted that one needed to correct for the larger body size that went with the large head, so that the non-Caucasian skull was downsized to fit the racial preconceptions. However, this correction was not made in comparing male with female skulls. Because females tend to have smaller bodies than males, their head sizes naturally would tend to be smaller. But taking into account the ratio of head size to body size would work against the preconception that the male brain is largest. So, where a correction factor was used for race, it was selectively not used for sex.

The attributions made by the left-hemisphere interpreter are especially important in determining human social perceptions of strangers who are different from us and fall outside our in-group identification. Cross-cultural research has shown that human beings make separate attributions about the stranger's degree of warmth versus coldness, on the one hand, and competence versus incompetence, on the other.[31] For example, the poor and the homeless are often judged as lacking in warmth and in competence, eliciting an emotional response of contempt. We are making a judgment about whether the stranger intends to do us harm (a friendly, sincere, or warm person is unlikely to cause harm, whereas a sinister, untrustworthy, or cold person just might). At the same time, the interpreter makes an attribution about whether the stranger is capable of causing harm, if that is his or her intent.

For members of our immediate in-group, as well as social groups that we aspire to join, the interpreter readily attributes competence and warmth.

The emotion triggered by members of these reference groups is admiration. Given the demographic breakdown of the United States, it is not surprising that, according to survey responses, the groups that the majority identify with and admire consist of middle-class, White, and Christian individuals.[32] The members of these groups are seen as good, admirable people. Notice that an attribution of competence is good for the in-group but bad for the out-group. In other words, the emotional reaction to competence is entirely different in the presence of an attribution of warmth (in-group members) compared with one of coldness (out-group members).

Outcasts of society, such as the homeless, are often seen as lacking both warmth and competence, eliciting feelings of contempt. Less vilified, because they have at least one dimension in common with the favoritism shown to the reference group, are the other two out-group prejudices. Some are seen as relatively high in warmth but low in competence. These groups, such as the elderly or the disabled, are stereotyped as benevolent and deserving of pity. Envy, on the other hand, is directed at those who are seen as threateningly competent and disliked. Envious prejudice has been directed at "nontraditional women, such as career women and feminists" and also "Asian Americans and other 'model minorities' stereotyped as excessively competent (too ambitious, too hardworking) and lacking sociability."[33]

The social attributions made about strangers have been investigated using functional magnetic resonance imaging (fMRI). A standard moral reasoning task was given to determine if in-group members, about whom attributions of warmth and competence are made, are treated differently than out-group members, who are seen as neither warm nor competent (e.g., the homeless). The task poses a moral dilemma: Is it acceptable to take the life of one person in order to save the lives of five others? The scenario presented read as follows. "An empty runaway streetcar speeds down the tracks toward five people. Joe, from an overpass, sees this accident unfolding. If Joe chooses, he can shove a bystander off the overpass to block the streetcar, saving the five people. How morally acceptable is it for Joe to push the bystander off the overpass?"[34] This problem was presented to American college students who had to make this decision about different types of individuals who would be sacrificed. For in-group members (i.e., students or Americans), there was reluctance to sacrifice even one person despite the fact that five would be saved. On the other hand, the results showed that cold and incompetent strangers, such as the homeless, were seen as more acceptable sacrifices. Taking the life of one homeless person was regarded as especially appropriate to save the lives of five others. Other less extreme out-groups of pitied individuals (e.g., warm but incompetent elderly) and envied individuals (e.g., the cold but competent rich) fell in between.[35]

The fMRI data revealed seven cortical regions with differential activity for sacrificing low warmth, low competence people in order to save high warmth, high competence individuals. All seven of them showed activation in the left hemisphere. Only two of the seven revealed bilateral activation.[36] In addition to these, the left middle occipital cortex showed more activation for sacrificing low competent compared with high competent victims. In a comparison of saving high competent individuals compared with the low competent, the left anterior cingulate was relatively more active. In short, the left hemisphere was heavily involved in the attributions made in this task, as anticipated by other research indicating the left hemisphere as the home of the interpreter. Attributions about others stem from the interpreter part of the modern mental ensemble. This, again, reflects how these neocortical networks color our experience of the social world.

Chapter 9

MORALITY

Mohandas Gandhi, the Mahatma, exemplifies for us the moral mind in his life of compassion for the poor and downtrodden and in his vigorous but nonviolent attacks on injustice through acts of civil disobedience. Throughout his life (1869–1948) Gandhi worked to free India from British colonial rule and to establish a universal moral standard for the treatment of all Indians, regardless of their caste, including British occupiers. Gandhi fought against British governmental policies, but only through non-violent resistance. His universal moral code mandated avoiding harm to other living beings and living with love for the good of others.[1] He put into action the Kantian categorical imperative: seek only those moral codes that ought to be universally applied to all. Fairness, in a word, was the driving principle of the moral reasoning exemplified in Gandhi's life. It was a sense of fairness shaped not only by compassionate caring but also by a need for justice:

> [Gandhi] was a tireless champion of the rights and well-being of the poor
> and powerless, not only in his own country but around the world. On a per-
> sonal level, he put himself in the service of others, often tending the sick
> and nursing the injured. . . . Gandhi's life is a story of compassion for those
> in need, a defense of universal justice, and a passionate articulation of the
> effectiveness of dialogue and nonviolent resistance. Each of these aspects of
> his moral reasoning demonstrates the depth and breadth of Gandhi's fair-
> ness. A close reading of how he addressed the major moral crises of his life
> also illustrates the typical integration of justice and care considerations in
> the resolution of real life moral problems.[2]

A concern for fairness arises from the capacity to reason about moral groups, taking into account what is ethically right. In perceiving events in the world, particularly the actions of human beings that impact other human beings, the interpreter comments on the degree of justice and caring manifested. The inner voice of those of us with a strong sense of fairness might say: "Everyone should get her fair share. It's wrong to *use* people. I wouldn't want to cheat anyone, any more than I would want to be cheated. I try to be kind to everyone. Everyone deserves respect. We're all in this together."[3]

One way to fairness is through sensitivity to social justice and equity. This path relies on the ability of human beings to employ logic and abstract concepts of equitable arrangements. It depends on the symbolic thought capabilities of human beings as well as on the advanced working memory needed to reason. Lawrence Kohlberg hypothesized three levels of development that must be passed through as children develop a sense of social justice.[4] In the first level, young children focus on the negative consequences of an action. Young children are highly egocentric, meaning that they see the world chiefly from their own perspective. An action is considered wrong precisely because it results in the child being punished for doing it. It is the parental authority figure who makes the decision about what is right or wrong. The next step on the path involves thinking about right and wrong in a social or interpersonal way rather than egocentrically. An action is wrong if it harms a friendship with another person or if it harms the family or the community. At this point, the child is able to think about ethically right actions being helpful to the group, not just the individual. The final step is to understand that, as members of society, we are obligated to obey universal rules of social justice. Understanding the concept of a social contract, and eventually coming to adhere to abstract principles of justice, completes the development of moral reasoning. Note that justice reasoning is heavily dependent on the human capacity for language. Only through language do we have a means for thinking about abstract concepts, including concepts like "justice" and "universal." Imagine trying to reason at this highest level entirely by drawing pictures rather than by talking or writing.

The alternative path to fairness places much greater emphasis on the power of empathy for the feelings of others. By emphasizing caring and compassion for others one can also reason about what is fair—this more emotional path to moral reasoning should be seen as a complement to the rational path of concern for the self, the group, and finally principles of social justice. This path to fairness thus exploits the advanced social intelligence of human beings. It relies more on being able to read the minds of others and feel their pain. However, an advanced social intelligence alone is not enough. Rather, the theory-of-mind capacities merely allow for the ability to engage in care reasoning, a form of reasoning that is just as dependent on an advanced system of working memory as justice reasoning. Similar to Kohlberg's three stages of moral reasoning, Carol Gilligan proposed three steps along the caring pathway as well.[5] Again, because children are egocentric, the first step involves self-love. Young children care for themselves only, orienting their concerns to individual survival. The second step requires moving beyond this selfish stage to a concern with the well-being of others. In fact, caring for the needs of others completely overtakes moral reasoning, as goodness is defined as self-sacrifice. Working memory must be combined with insight into the feelings of others to achieve this concern. The self-sacrificial stage is transcended when people take the final step, whereby the needs of others and the self are morally equated. The focus here is on a morality of nonviolence in which one's actions should neither harm the self nor harm others, as seen in the protest practices advocated by Gandhi. A refusal to hurt others thus leads one to fairness, but it does so through the development of caring and compassion rather than through abstract principles of social justice.

In the most extreme form of altruism, there is a willingness to sacrifice even one's life in order to save the lives of others. This could be understood in terms of the second stage of care reasoning in which taking care of the needs of others above all else is the standard of what is right and good. However, it could also be, in the case of life or death circumstances, that the injunction to do no harm to others or to oneself boils down to a choice. Do I harm myself or another? In natural disasters, ship or airplane accidents, and the combat of war, grave and agonizingly real moral dilemmas must be confronted. According to the perspective advanced here, making the altruistic decision of self-sacrifice ought to be more common when the ensemble of the modern mind becomes engaged to overcome the instinctual response of self-survival. Unless our capacity for advanced working memory and social intelligence can be brought to bear, the most likely response is the fast and habitual one of self-preservation.

A comparison of two famous disasters at sea makes this point well.[6] On the night of April 14, 1912, the RMS *Titanic*, the greatest luxury ship of its day, struck an iceberg in the North Atlantic on its maiden voyage from South Hampton to New York. Of the 2,207 confirmed passengers and crew, 1,517 were lost to the frigid sea. Over a period of nearly three hours, between the time the ship struck the iceberg and the eventual sinking, numerous acts of altruism occurred. There were not enough lifeboats for all on board, and the ship's crew called for saving the women and children first. Hundreds of men perished rather than selfishly taking a place in a lifeboat, despite the certainty of a painful death for those who remained with the ship. Compared with a reference group of older adults (greater than thirty-five years of age) who traveled third class and had no children, significant advantages in survival rates were found for children under the age of sixteen, women between the ages of sixteen and thirty-five, and parents with children. Thus, the social injunction to save women and children first in fact resulted in altruistic acts of self-sacrifice on the *Titanic*.

On May 7, 1915, the cruise liner RMS *Lusitania* was torpedoed by a German U-boat just a few miles off the coast of Ireland. The ship was similar in design to the RMS *Titanic*, and their passenger lists and crews were comparable. However, the situation facing the passengers on the two ships was different in a critical way. The *Lusitania* sank in a scarce eighteen minutes. With so little time, there was little chance for reflection by those on board. There was no time to engage the deliberative processes made possible through our advanced social intelligence and working memory capacity. Of the 1,949 confirmed passengers and crew on board, 1,313 died. There is no reason to believe that the men on the *Lusitania* were less altruistic in general than those on the *Titanic*. Nor is there reason to believe that caring for the lives of women or children ahead of their own was not stressed. In fact, "in both disasters, the captain issued orders to their officers and crew to follow the social norm of 'women and children first.' These orders were successfully carried out on the *Titanic*, but not on the *Lusitania*, due to the time constraints and problems launching the lifeboats."[7] Relative to the reference group, children under the age of sixteen, females sixteen to thirty-five years of age, and parents with children had no advantages in survival. For males sixteen to thirty-five years of age in the prime of physical condition, survival probability showed a small but statistically significant advantage over the reference group of older, childless, third-class passengers on the *Lusitania*. In sinking in just eighteen minutes, such an advantage would be expected. Yet there was no such advantage for this physically able group of males on the *Titanic*. In fact, young males had a worse survival rate than the reference group.

Fairness will not rule human actions with respect to the rights of others if a person fails to engage in the deliberation required by justice reasoning or care reasoning. Lack of time is only one reason why such failures occur. Another is a failure to devote the executive attention required by the effortful process of moral engagement. When an authority figure orders one to commit an immoral or unethical act, it requires considerable mental effort to think through what is right and what is wrong. Whether one arrives at fairness by appealing to universal principles of social justice or by asserting a priority of nonviolence to the self and others, reflection is required. None of this will happen if the individual simply disengages from the process. Such moral disengagement is the easy path to take when the blame for wrong actions can be placed elsewhere, when one is simply following orders. Even evil acts can be mindlessly committed if the responsibility is displaced to the authority figure. Albert Bandura explained why: "Moral control operates most strongly when people acknowledge that they cause harm by their detrimental actions. . . . Disengagement . . . operates by obscuring or minimizing the agentive role in the harm one causes. People will behave in ways they typically repudiate if a legitimate authority accepts responsibility for the effects of their conduct."[8]

Stanley Milgram conducted extensive research on how obedience to an authority figure disengages the human faculty for moral reasoning. He found that people were willing to administer electric shocks to another human being when instructed to do so by an experimenter in a psychology research study. Compliance was common even when the shocks administered were extremely dangerous (450 volts). The import of this finding is immense, as Jerry Burger noted in his more recent attempt to replicate the results: "the haunting images of participants administering electric shocks and the implications of the find-ings for understanding seemingly inexplicable events such as the Holocaust and Abu Ghraib have kept the research alive for four decades."[9]

In Milgram's most famous experiment, he created a situation in which a research participant believed he was teaching a partner the correct answers in a paired-associate learning task. Each stimulus word was paired with a correct response word. If the learner made a mistake in responding to a stimulus word, the experimenter—wearing a white lab coat to indicate his status as an authority figure—told the participant to deliver a punishment to the learner in the form of an electric shock. By punishing incorrect responses, the participant was actively teaching the learner the right answers. The shocks ranged from 15 to 450 volts, from mild to extremely dangerous. The teachers were further instructed to start with the 15 volt shock and then progress up the scale of 30 switches.

In reality, the learner received no shocks, since the learner was in fact a confederate of the experimenter. In the original experiment, the learner was hidden from view of the participant, but the sounds of his cries of pain to the shocks could be readily heard at the 150-volt level. The learner complained that his heart was bothering him and that he wanted out. From 150 to 330 volts, the learner yelled in pain and insisted on stopping the experiment. After 330 volts, the learner could no longer scream of protest, implying he was physically injured by the teacher's infliction of punishments. In each case, the experimenter sat a few feet from the teacher and repeated the command to continue raising the shock level with each mistake whenever the teacher showed signs of wanting to discontinue. The experimenter did not end the session until there had been four consecutive trials in which the teacher had resisted administering a shock or had administered the full 450-volt shock.

Jerry Burger tested a sample of the general public in 2006, and 70 percent of subjects continued beyond the switch for 150 volts when they first heard the leaner protest, only slightly less than the 82 percent of those tested in 1961 and 1962.[10] Milgram reported that 65 percent of the volunteers in his original experiment complied with orders all the way to the top of the scale, the highly dangerous shock of 450 volts. To avoid the ethical problem of ordering participants to go that far, in the contemporary replication study par-ticipants were stopped if they attempted to exceed 150 volts. Still, it was possible to extrapolate how many likely would have continued to the top of the scale. "Because 79% of Milgram's participants who went past this point continued to the end of the shock generator's range, reasonable estimates could be made about what the present participants would have done if allowed to continue. Obedience rates in the 2006 replication were only slightly lower than those Milgram found 45 years ago."[11] The power of the situation—an authority figure issuing orders to obey—can encourage us to skip the deliberative process needed to do the right thing. Although human beings are endowed with the modern mental ensemble that can lead to moral choices, a failure to engage these resources renders them moot.

The abstract reasoning processes enabled by our advanced working memory can be elicited in laboratory tasks that pose a moral dilemma, allowing the brain networks to be identified. Would you activate a switch that diverts the path of train so as to save five people if it meant killing one? Most people take a utilitarian view that the lives of five people outweigh the life of just one, so they would in fact divert the train to the track with only one person on it. Neuroimaging data revealed that this impersonal moral dilemma activated areas in the right middle frontal gyrus and the posterior parietal cortex associated with working memory and reasoning tasks in general.[12] These regions were affected no differently in the impersonal moral dilemma and a nonmoral decision task. However, by switching the task instructions just slightly to make the decision personal, the emotional areas of the brain were instead highly activated. Would you push a stranger off a bridge to save the lives of five others? The stranger will die as a result of your decision, but the stranger's body on the tracks will stop the train and thus save five others. In this scenario, bilateral activation of the medial frontal gyrus, the posterior cingulate gyrus, the angular gyrus predominated.[13] Not only are these regions involved in emotional processing, they also constitute the default network underlying mental time travel.[14] Imagining the future or recollecting the past requires the brain to simulate an experience that is not present here and now. Possibly the same kind of simulation must transpire before one can make a personal moral choice that is emotionally laden. One must actually imagine pushing the person off the bridge before a decision can be reached. By contrast, the impersonal moral dilemma can be tackled as a problem of cold reason.

MORAL INTUITIONS

Although reasoning on the basis of justice or care is fundamentally important to human morality, it is not the only basis for behavior. Moral intuitions that operate quickly, automatically, and without rational deliberation can also control behavior. Although deliberative reasoning is commonplace, it may be used by people "to seek evidence in support of their initial intuition and also to resolve those rare but difficult cases when multiple intuitions are possible."[15] Jonathon Haidt makes the point that our intuitive moral judgments are tapped so quickly that they can bias the course of more deliberative reasoning.

Offensive foods, such as rotten meat, elicit strong feelings of disgust automatically. Disgust is readily elicited by pictures of "slimy, moist surfaces or colors reminiscent of body fluids" because they indicate the presence of pathogens that are dangerous to human health.[16] In a different way, disgust can also be powerfully elicited by morally taboo behaviors such as incest. It is immediately obvious to people that, say, having sex with a brother or sister is disgusting. Although the brain shows a common pathway in registering disgust to these two scenarios, they also differ in several ways. From neuroimaging data, researchers conclude that "despite their tendency to elicit similar ratings of moral disapproval, incest-related acts and non-sexual immoral acts entrain different, but overlapping brain networks."[17]

In these circumstances, the interpreter of the left hemisphere engages in the moral domain as it does in all domains. It tries to provide an explanation. An automatic intuition yields a judgment that the interpreter then tries to account for using all that is known about the current situation. For example, in making a moral judgment about a harmful act it is important to know the intention of the perpetrator. Did a man who shot his hunting partner do so by accident or on purpose? There is an obvious difference in our reprehension of the shooter depending on how we read his mind. The critical feature in our legal code between manslaughter and murder is whether a death results by accident or by intent. The advanced social intelligence of the human brain includes a network in the right inferior frontal cortex that aids this theory of mind skill.[18] In split-brain patients, the computations made in the right hemisphere about the intentions of an actor are not transmitted across the corpus callosum as normally would occur. Thus, the left hemisphere of the split-brain patient cannot readily take into account the intentions of others in judging the moral probity of their actions.[19] Indeed, for split-brain patients the outcome—the perceived act—is all that matters in deciding right or wrong. An action is immoral if it led to a bad outcome regardless of whether the actor had no intention to harm. When asked why an innocent act that accidently caused harm is morally offensive, the left hemisphere of the split-brain patient concocts an explanation. As Michael Gazzaniga concluded from his studies on moral judgments in split brain patients, "the interpreter introduces a misleading level of certainty about the reasons that moral judgments are made. Yet the narrative of the interpreter helps us make sense of our social environment."[20]

In normal individuals with an intact corpus callosum one can see the same effect through the use of hypnosis. The basic idea was to implant a posthypnotic suggestion that would elicit a strong feeling of disgust whenever a particular arbitrary word was read (i.e., *often*). For highly hypnotizable individuals, it is possible to suggest that the individual feel something or do something later on when they come out of hypnosis, and they will reliably do so. Importantly, they will do so without any clear understanding of why. That is, they may not have a memory of having been told to feel disgust when reading *often*, but they will nonetheless feel disgust. The participants were presented with a variety of scenarios that varied in their offensiveness. At the high end was engaging in incest with a cousin. Bribery was not particularly disgusting, but it was viewed as morally wrong. The result showed that if the trigger word, such as *often* appeared in the description of the scenario, it heightened the feelings of disgust and the severity of judgments that the scenario is morally wrong. This outcome held for both bribery and incest.[21]

Those in the experiment who were not aware that the posthypnotic suggestion was having this effect on their emotions and moral judgments were in much the same situation as the split-brain patients; or, in the shoes of us all when disgust is immediately and automatically registered before any reasoning can take place. In these circumstances, the left-brain interpreter plays a critical role of trying to account for why we feel and judge the way we do. In another hypnosis experiment, a perfectly innocent scenario was associated with the trigger word.[22] It described a student council representative who was put in charge of organizing discussions around topics of interest to both the students and faculty. Without the presence of the hypnotic trigger word, this story elicited neither disgust nor moral concerns. Yet with the trigger word, the student's actions were seen as nine times as disgusting and five times as morally wrong. To justify their odd reactions, the participants made up explanations: "It just seems like he's up to something," or he is just "a popularity-seeking snob."[23] These views were not expressed when there was no posthypnotic trigger to feel disgust. As Michael Gazzaniga noted, "This behavior suggests that automatic social evaluation produces a judgment, which the interpreter then registers and attempts to explain."[24]

EMPATHY

Human beings have an acutely developed capacity to comprehend and tend to the emotional experiences of others. If we observe sadness in another person, then we, too, can experience sadness; in the same way, happiness can be contagious and spread in a social group simply from observing the happiness of another. At a basic level of perception and action, empathy for the feelings of others involves the mimicry of facial expressions. We reflexively frown or smile to reflect the sadness or happiness seen in another. Yet empathy involves more than this perceptual-motor response to the feelings of others. It further invokes the high level of social intelligence that allows us to understand others as intentional agents with minds of their own. The rich set of theory-of-mind developments in early childhood build upon the more primitive emotional mimicry networks. Empathy further draws on the executive functions of our advanced working memory to take the perspective of another person and see the world from his or her eyes. Human empathy, then, represents a complex interaction of neural capacities that come together in a radically different way from other species.

A precursor to human empathy definitely has been found. The premotor cortex of primates possesses a special type of neurons, called mirror neurons because they appear to allow mimicry. In a macaque monkey, recordings of the firings of these neurons can be observed when the monkey grasps or manipulates an object, but also when another monkey, or even a human experimenter, performs the same action.[25] The mirror neurons in the ventral premotor cortex are part of a circuit that extends from the posterior parietal cortex, passing through the posterior superior temporal cortex. In seeing an action performed, the motor neurons that produce the same action are stimulated via this circuit.[26] This requires a common code that couples perception and action. The way the brain codes for a particular action entails a representation of the perceivable effects of that action. In the case of perceiving an emotional expression of another monkey, the codes would be activated for producing that same expression and mimicking it. The mirror neurons and the network linking perception with action can thus provide a foundation for empathy, in the sense that one can catch the emotion experienced by another. Emotion contagion can be seen in human beings who viewed pictures of happy or angry faces for a very brief time, too fast for their conscious registration. Nevertheless, measurements of their facial muscles showed they were mimicking the happy or angry expressions presented. "Furthermore, this effect was stronger for the participants who scored higher on self-reports of empathy."[27]

Psychologists have recognized various facets of empathy, starting with the fact that feelings are elicited by another person's emotional state.[28] This can result in a sharing of the emotional state of another person, although it is also possible to empathize with another without sharing in the same emotion. A distance can be maintained, while still recognizing what the other person is feeling. Actually sharing in the emotional experience of another can occur in different ways. On the one hand, the emotional expressions of another person can automatically trigger the same expressions in us through mimicry. On the other hand, we can devote executive attention to imagining how the other person is feeling and thinking, in the metaphoric senses of stepping into their shoes or seeing the world with their eyes. Lastly, there is the self-regulation that underlies our ability to keep our distance, while still empathizing. Complete sharing of another's negative emotional state of, say, sadness, and complete adoption of another's point of view, could be personally distressing and even paralyzing. We would not be able to respond effectively to the emotional needs of others if we ourselves were to become too sad and depressed to act.

Regions in the prefrontal cortex mediate the executive function of taking another person's perspective. Imagine a situation that would induce the emotion of shame. Now, imagine how you would feel in that situation? Finally, imagine how your mother would feel in the same situation. A neuroimaging study compared the shame-inducing situation with a control situation that was emotionally neutral. "Regardless of the affective content of the situations depicted, when the participants activated their mothers' perspective, activation was detected in the frontopolar, ventromedial prefrontal, and medial prefrontal cortex, and the right inferior parietal lobule—congruent with the role of these regions in executive functions associated with the perspective taking process."[29] The theory-of-mind capabilities of our advanced social intelligence must be leveraged by the executive attention of our advanced working memory to pull off the feat of taking another person's perspective.

One study used functional magnetic resonance imaging (fMRI) to detect whether we literally can feel another's pain. Pictures of people with their hands or feet in painful situations were shown to research participants who were asked to rate the level of pain they vicariously experienced from viewing the pictures. When instructed to imagine themselves in this situation, participants reported feeling high levels of pain, and the fMRI results revealed that the anterior cingulate cortex and anterior insula were activated, just as they are when pain is actually experienced firsthand.[30] The higher the activation of these areas, the more pain the participant reported would be produced. Similarly, these same regions fired when a participant received a painful stimulus. But they also fired when they then simply observed a partner receiving the same stimulus—the brain network was registering pain felt by another person, a precondition for empathy.

The mirror neurons that enable the brain to recognize an action appear to capitalize on our uniquely human theory of mind. Because human beings are capable of understanding others as intentional agents, it is possible to recognize the intention behind the action. Suppose you observe a hand grasping a cup. Is the actor's intention to take a drink from the cup or to wash it? If the context shows the cup with a teapot, a plate of cookies, and a jar arranged for having tea, the action would then be linked to the intention of drinking. By contrast, if the context shows the cookies mostly eaten, the lid off the jar, and a soiled napkin under the teapot, then the action would be linked to the intention of washing. Using fMRI, researchers measured the response of the brain to the context alone, the isolated action, and the action in context.[31] By comparing the neural activity generated by each of these conditions, they were able to isolate a region in the right inferior prefrontal cortex—a region known for the presence of mirror neurons—that specifically recognizes the intention. For example, subtracting the activation found for the isolated action from the action plus, say, the washing context, the activated region still remaining would be the site for identifying the intention as opposed to the action.

A final facet of human empathy is the ability to regulate one's emotional response. When emotions are induced simply by watching another person experience pain or express sadness, the emotions must be modulated in order to feel empathy for the person instead of just getting caught up in the same plight. Part of this process of keeping our distance from the object of our empathy activates a region in the right hemisphere at the junction of the temporal lobe and the parietal lobe.[32] The region is associated with self-agency. Another part explicitly aims to downplay the significance of the emotion "by denial of relevance (i.e., taking a detached-observer position) by generating an image of the observing self unaffected by the target . . . to reduce the subjective experience of anxiety, sympathetic arousal, and pain reactivity."[33]

MORAL MUSCLE

At its root, morality requires a standard of proper conduct. Knowledge of good versus evil is part of our nature. If it were not, then moral judgment and choosing to do what is right rather than wrong would not be possible. But simply the knowledge of right and wrong is by itself insufficient for moral action. One must also be able to inhibit evil thoughts and not act upon them. Is not a person morally accountable for their actions if they know the difference between right and wrong but cannot avoid acting on an impulse to do wrong? Understanding fairness through moral reasoning is only part of the story. The ability to monitor one's behavior relative to a standard of conduct and to alter one's behavior to bring it in line with the standard is referred to by Roy Baumeister as "moral muscle."[34]

Just as precursors were found for language in nonhuman primates, the mirror neurons that link perception and action provide a foundation for the full scope of human empathy and morality. As already shown, human empathy builds on this foundation, adding advanced social intelligence and the human executive functions, which combined permit the reading of others' minds and the taking of their perspectives. But another part of the modern mental ensemble also contributes. Our moral intuitions about what is right and what is wrong were passed down through the medium of language from one generation to the next for thousands of years in the oral traditions of countless cultures. Eventually, the socially transmitted moral codes became the basis for the written legal codes of Neolithic humans. Beginning with Babylon's Code of Hammurabi in the eighteenth century BCE, cultural evolution fueled by written language has brought us, thirty-nine centuries later, the legal systems of today. Contemporary national and international law reflect the confluence of language and morality that sculpted human cultures and histories.

A culture's standards of moral conduct often derive from its religious doctrines. The Ten Commandments of Judeo-Christian cultures reflect an admixture of theology and morality. Spoken by God on Mount Sinai (Exodus 20:3–17), the commandments provide rules for both humanity's vertical relation with God (e.g., "You shall have no other Gods before Me") and the horizontal relations among people (e.g., "You shall not murder"). In addition to proscribing murder, God commands against committing adultery, stealing, bearing false witness against one's neighbor, and coveting anything that belongs to one's neighbor.

It was Christian theologians in the Middle Ages who delineated the Seven Deadly Sins. In each case, a clear standard of moral conduct was articulated, and human failures to conform to the standard were viewed as sins against God's will for humanity. Gluttony, for example, is an overindulgence of the human appetite for food or drink. Although the religious context of gluttony has faded in contemporary society, overeating and overdrinking are as popular as ever. The US Centers for Disease Control and Prevention reported that the prevalence of obesity in the United States rose markedly during the last decades of the twentieth century, with the most recent figures indicating that 35.7 percent of the adults over the age of twenty were obese in 2009–2010.[35] Plentiful sources of energy-rich food, a largely sedentary lifestyle, and excessive consumption have resulted in more than a third of the population meeting the body-mass-index definition of obesity. Neither greed nor lust are endangered either. The recent collapse of capital markets throughout much of the Western world in 2008 has painfully reminded us that greed reigns. However, it would be mistaken to think that greed in the twenty-first century is any more intense than in any past century. The crash resulted at least in part from an overemphasis on short-term profits that encouraged excessive risk.[36] Surely, no one could seriously argue that human beings are more lustful today than they have ever been. Yet it is now much easier to document those lustful thoughts simply by recording the volume of Internet searches for pornography. Natasha Vargas-Cooper, writing in 2011 in the *Atlantic*, concluded that "pornography is now, indisputably, omnipresent: in 2007, a quarter of all Internet searches were related to pornography. Nielsen ratings showed that in January 2010, more than a quarter of Internet users in the United States, almost 60 million people, visited a pornographic Web site."[37]

Sloth, wrath, envy, and pride round out the list of seven. It is safe to say that none are in any danger of extinction. The same kinds of failures of self-control described by theologians in the Middle Ages are, of course, very much part of human life today. The central point is that for human beings to conduct their lives morally, they must have knowledge of a set of standards. If societies tolerate or even condone gluttony, greed, and lust, then there is no basis for expecting people to exercise self-control. People in each generation assume the obligation of transmitting standards of conduct to their children through oral instruction. The written word, in sacred texts, in theological commentaries, in the law and in judicial records, not to mention in the advice columns of countless newspapers and magazines, also provides knowledge of the standards. Today, no less than in the Middle Ages, the concept of what is proper and fair can be readily obtained chiefly because of language. It is hard to imagine how human morality could have ever developed and endured without language to encode and store the standards.

Mental time travel also is central to the establishment of moral standards. It frees human beings from the present moment by reconstructing past events and imagining future ones, and it is therefore responsible for much good. Perhaps this ability to think prospectively as well as retrospectively enabled early humans to conceptualize the notion of time itself. It is foresight in particular that has a role in determining the codes of right and wrong. Thomas Suddendorf and Michael Corballis explained that

> humans can forecast the outcomes and choose to act now to secure future needs. Cultures have evolved complex moral systems that judge actions as right or wrong based on what the actor could or could not have reasonably foreseen to be the future consequences of the act. Law, education, religion, and many other fundamental aspects of human culture are deeply dependent on our ability to reconstruct past and imagine future events.[38]

Knowing a moral standard is of little use if one fails to monitor one's own behavior and compare it with the relevant standard. For this reason, a virtuous individual is also necessarily a self-aware individual. Large crowds of people can create a feeling of deindividuation or a loss of self-identity. Self-monitoring of behavior fades away as the individual melds with the larger group. If the group is inclined to destructive behavior, its actions can escalate out of control. Acts of violence, vandalism, and theft can erupt from a mob far more readily than from any one individual, with the exception perhaps of a psychopath. In a sense, a mob can become a psychopath. As another example, alcohol intoxication impairs self-awareness and "alcohol is well known to be implicated in a broad range of nonvirtuous behaviors ranging from interpersonal violence to sexual misdeeds.[39]

Finally, virtuous behavior depends on the self-control of impulsive thoughts and feelings. Transient feelings of anger or jealousy can, if acted upon, produce verbal or even physical violence. Having moral standards and being self-aware are by themselves insufficient without self-regulation. It is here that the concept of moral muscle makes good sense, for as with any muscle, fatigue can deplete one's capacity for moral conduct. Human beings are vulnerable to what Baumeister calls ego depletion.[40] Tasks that require sustained attention and mental effort are depleting, just as is sustained physical exertion. Exercising self-control over impulsive thoughts and feelings is one of the tasks that takes mental effort and causes ego depletion. The executive function of inhibiting unwanted thoughts can fail as a result of mental fatigue. The mental work of constraining one impulse can make it harder to resist a different, wholly unrelated impulse. Executive attention is a limited resource that can be overloaded, causing problems in self-regulation. As Baumeister puts it, "in all acts of volition the self uses some resource that operates like an energy or strength, and after such an act the self's stock of this resource is depleted."[41] After a hard day of work, it is especially difficult to regulate one's behavior and stick to goals. For the dieter, for example, the temptation of a large dinner or a high-calorie dessert is hard to resist after exerting self-control accomplishing tasks at work all day.

CRIMINAL RESPONSIBILITY

Roy Baumeister noted that the loss of self-control under ego depletion could account for lifestyles that characterize those repeatedly in trouble with the law. For example, most criminals are arrested repeatedly but for different crimes, contrary to the view of criminality as a specialized career choice (as movies like to portray it). Moreover, most criminals tend to show patterns that reflect poor self-control even in legal activities: "For example, criminals are more likely than others to smoke cigarettes, drink alcohol, be involved in unplanned pregnancies, and have erratic attendance records at school or work."[42]

Does lack of capacity to self-regulate one's behavior imply a lesser degree of moral accountability for wrongdoing? The law has long recognized the criticality of *mens rea*, or guilty mind. If someone commits a crime without the capability of understanding the consequences of his or her actions, he or she lacks criminal intent and should not be punished to the same extent as one who does possess this capacity. Here is a clear and noble example of moral reasoning enshrined in the rule of law. Yet it is far less clear in what circumstances this principle applies.

In their article on neurolaw, Annabelle Belcher and Walter Sinnott-Armstrong reviewed the facts of a notable case on criminal responsibility and brain research.[43] In 1993, a seventeen-year-old male by the name of Simmons plotted, fully premeditated, to murder a woman named Shirley Crook. He brought with him two younger friends, Tessmer and Benjamin. His plan involved burglarizing the victim, followed by abduction and then murder to dispose of the evidence. She was bound, gagged, and thrown off a bridge in a state park near St. Louis. She drowned that day in the waters of the Meramec River. Charged with first-degree murder, Simmons was convicted, not surprisingly given that he confessed to the murder and performed a videotaped reenactment at the crime scene. In addition, Tessmer testified against him, documenting his plotting of the crime and his bragging later about his deeds. Under Missouri law, the jury recommended a death sentence, and it was so imposed by the court.

For the next eleven years, appeals to the death penalty decision were upheld, but the decision was nevertheless brought before the US Supreme Court. Although all the elements of first-degree murder applied to Simmons's actions, and although Missouri law allowed for the death penalty in such a case, the Court heard the case because the defendant was not an adult at the time of the crime. In its 2005 *Roper vs. Simmons* decision, the Court concluded that the immaturity of adolescents relative to adults reduced the murderer's criminal responsibility. Based on the Eighth and Fourteenth Amendments to the US Constitution, the Court ruled that capital punishment for crimes committed under the age of eighteen years, regardless of their heinous nature, was forbidden. Put differently, society has a moral responsibility to treat the actions of a seventeen year old differently than it treats the actions of an adult. With the 5–4 vote, the Court effectively changed the age of full responsibility for premeditated murder from sixteen to eighteen years of age.

The majority opinion cited three critical differences between juveniles and adults.[44] Juveniles lack maturity and so are more inclined to reckless action, show vulnerability to external influences such as peer pressure, and continue to develop in their moral judgment as their character becomes more fully formed. Several *amicus* briefs had been filed documenting the neuroscience of adolescent brain development, although the justices did not directly cite such evidence in their opinion. Three facts of developmental neuroscience stand out as especially relevant. First, the limbic system that mediates reward processing is affected by hormonal changes during puberty. It becomes especially sensitive to social and emotional stimuli, contributing in part to psychosocial immaturity. As Lawrence Steinberg has explained, for teenagers "the mere presence of peers makes the rewarding aspects of risky situations more salient by activating the same circuitry that is activated by exposure to nonsocial rewards when individuals are alone."[45] This aspect of developmental change could explain why adolescent risk taking is so closely tied to being in groups. Second, the executive functions of working memory, on the other hand, are mediated to a large extent by prefrontal cortical regions that mature slowly throughout adolescence and young adulthood.[46] The juvenile's ability to inhibit impulses, plan ahead, resolve conflicting goals, and juggle multiple streams of information is still developing as the prefrontal cortex matures. This also contributes to psychosocial immaturity because the executive control network has difficulties in regulating risky behaviors when the socioemotional network is highly active in the presence of peers. Third, logical reasoning ability in the sixteen year old is equivalent to adult levels.[47] In the absence of peer influence and socioemotional factors, adolescents can make competent decisions. However, what would otherwise be competent decision making based solely on the teenager's logical abilities is often compromised by psychosocial immaturity.

Besides slow brain maturation, brain injuries raise another question of relative moral culpability. Damage to the ventromedial prefrontal cortex yields a distinctive pattern of moral judgments that appear devoid of the emotional empathy typical of normal human brain functioning. Consider the standard moral dilemma in which five people will be saved but one must be sacrificed. In the impersonal situation, this is achieved by deciding to activate a switch that diverts a train away from the five, but regrettably toward a sixth person. The personal situation has exactly the same cost-benefit ratio of saving five and losing one, but it imposes an emotional cost of personal, physical involvement, namely, pushing one person off a bridge to save five others. The first problem lends itself to a cold, rational calculation that saving five is worth the loss of one life. It is what moral philosophers refer to as a utilitarian moral judgment, in that there is greater utility in intervening to divert the train. However, in the normal population, fewer people are willing to push someone off a bridge to achieve the same effect. Patients with lesions to the ventromedial prefrontal cortex and orbitofrontal cortex, however, make no such distinction.[48] It is as if the emotional repugnance of physically pushing someone to his or her death does not enter into their moral calculation, and they remain utilitarian even in the personal situation.

People diagnosed with antisocial personality disorder (APD) show a pervasive disregard for social norms and willfully violate the rights of others. The term psychopath is sometimes used as an alternative to APD when stressing the individual's impulsiveness, manipulative deception, self-centered narcissism, and complete lack of both empathy for the feelings of others and guilt for violating them. A human being who willfully violates others without feeling empathy or guilt is capable of serious violence. Not surprisingly, perhaps, the incidence of APD is ten times more likely in prison populations than in the general population as a whole.[49]

It could be that antisocial acts of violence can be related to an enduring trait of personality, and, in some cases, there may be an underlying problem in the brain. Even so, it is important to recognize that brain injury is not necessary for a diagnosis of APD or psychopathy. Nor are these conditions defined by violence per se, for not all antisocial acts involve aggression. Rather, brain injuries to the prefrontal cortex might best be seen as a sufficient but not necessary cause of the violence that is found in cases of APD and psychopathy.

With these facts in mind, then, should a seventeen-year-old be held fully accountable for the act of premeditated murder? How about a psychopathic individual with a known brain injury that would impair prefrontal executive functions? These are clearly questions for our societal institutions of law that can only partially be informed by the findings of neuropsychology. In using neuroscience in the legal arena, it is important to keep in mind its limitations.

With regard to brain maturation, in most US states an individual can be tried as an adult at the age of fourteen, but the purchase of alcohol is usually restricted to those at least twenty-one years of age. An eighteen-year-old can serve in the military and vote, whereas a sixteen-year-old in most states can drive. A society draws these arbitrary lines depending on its assumptions about the moral agency of adolescents—which responsibilities can be handled and which cannot. An act of murder cannot be tolerated by society regardless of the brain abnormalities that may underlie the act. Whereas the nature of the punishment imposed might be affected by the neuroscience of the case, the law itself is a societal compact that cannot simply excuse the behavior, even if neuroimaging were able to tell us everything we needed to know about the potential danger posed by immature or injured human beings.

Moreover, the development of psychosocial maturity is a continuous process throughout adolescence. There is no single moment when a person's brain is "mature enough." The law must define boundaries on an inherent continuum. There are also large individual differences in the development of psychosocial maturity. A scientist can only say, in regard to a specific case, that a fifteen-year-old would generally be less mature than an adult. Whether a specific fifteen-year-old is more or less mature than a randomly chosen twenty-five-year-old is not a statement that science can make with certainty. The fifteen-year-old could be another Mahatma, whereas the twenty-five-year-old old could be a psychopath capable of cold-blooded murder without remorse. Similarly, in the case of brain lesions, there are wide variations in the location and amount of damage. It is highly individualized. How can one say with certainty how much damage, and in what location, exonerates one from responsibility? In short, individuals must still exercise their capacity for moral reasoning to arrive at a just and fair decision. The findings of neuroscience can inform such a decision, but they cannot take its place.

Chapter 10

SPIRITUALITY

T he oldest known temple appears to be Gobekli Tepe, located in Turkey not far from the ancient city of Urfa.[1] It consists of large rings of stone pillars, standing as tall as sixteen feet, with the largest ring measuring sixty-five feet across. The pillars weigh between seven and ten tons. Many know of the familiar megaliths of Stonehenge, which date from about five thousand years ago in what is today England. However, the astonishing discovery of Gobekli Tepe moves back the earliest known holy place of religious worship to eleven thousand years ago. This Mesolithic date preceded the origin of metal tools and pottery. As with other megaliths, its purpose appears to be symbolic and related to abstract beliefs rather than some pressing survival need. What deep human need would motivate a prehistoric people to labor so long and intensely on such structures? Besides the time and effort required to erect the pillars, many are adorned with intricate carvings of foxes, lions, scorpions, and vultures.

The existence of a Mesolithic temple raises the possibility that spirituality can be traced further back to the modern humans of the Upper Paleolithic. Scholars have proposed before that religion, "among the most powerful of all social forces," has been "here as long as there have been human beings (e.g., it has been suggested that humans be thought of as *Homo religiosus* because religion has been present as long as there have been *Homo sapiens*."[2] The novel proposal advanced here is that human spirituality *necessarily* dates at least to the Upper Paleolithic because it is an emergent and inevitable outcome of the modern ensemble of mental parts discussed throughout this book.

With emotion and social relationships, human beings plainly share much in common with other species. Even with regard to morality, some scholars see signs of the rudiments of empathy in the great apes, and certainly the neural substrate of mirror neurons is found in primates in general. Arguably, the full meaning of the term moral mind—as it has been amplified by the modern mental ensemble—is only applicable in *Homo sapiens*, but there is not a complete discontinuity with nonhumans. By contrast, spirituality would seem to be quintessentially human and only human. Only human beings inquire as to their reason for being, their purpose in living, and their destiny after death. Only human beings can conceive of the prime mover that unfolded the universe from nothing and ponder the mystery of existence.

Here it will be argued that speech—that is to say, the capacity for symbolic thought and language—is only one part of the ensemble that underpins human spirituality. Equally important is an advanced system of working memory that allows us to use language in both scientific and theological reasoning about highly abstract concepts. The ability to mentally envision our own death is also central to the religious concern with being—and not being. The nature of being—why the world exists at all instead of the alternative of nonexistence—is of course a foundational question of philosophy that lies, too, at the heart of religion. Without a left-hemisphere interpreter seeking causal explanations for our perceptions, recollections, and fantasies, there would be neither theological inquiry nor scientific inquiry. The interpreter that enables us to infer hidden causes and explain reality drives thinking about questions open to scientific investigation and those that fall into the province of theology. The scientific question of how the heavens go is of interest, but so is the theological question of how we go to heaven.

In short, human spiritual aspirations and the religions that follow from them are a necessary and inevitable consequence of the modern ensemble of mind. Theological inquiry emerges from a mind with symbolic thought and language, aided by an advanced working memory for reflecting on ideas, and abetted by advanced social intelligence for finding common ground. In a species with the capacity to infer hidden causes and interpret perceived events, it is inevitable that we will inquire about the first cause that began the universe. Yet the use of mental time travel to anticipate our own death and possible nonexistence heightens the urgency of these quests for theological understanding. Why are we here? Is there a purpose and plan in our existence? Where are we going after death?

The argument here extends a position laid down more than a century ago by William James. James gave the prestigious Gifford Lectures at the University of Edinburgh and published his remarks in 1902 in *The Varieties of Religious Experience.* In his celebrated book, James explored the distinctive ways that people express their spirituality while making clear that all such varieties reflect an inherent and unavoidable facet of human psychology. Coming to terms with death is only part of religion's function. At its core, religion grapples with the essential mystery of our existence. It is our life on this planet in an incomprehensibly vast universe that begs for answers as much as our death. Although we can avoid thinking about it, the mystery of our being in the first place—the inevitability of not knowing for certain—confronts us all as a fact of being human. In her contemporary radio series *Speaking of Faith*, Krista Tippett expressed this point well in what follows.

> Mystery is the crux of religion that is almost always missing in our public expressions of religion. It eludes and evaporates beneath the demeaning glibness of debates and sound bites. If mystery is real, even more real than what we can touch with our five senses, uncertainty and ambiguity are blessed. We have to live with that, and struggle with its implications together. Mystery acknowledged, is, paradoxically humanizing.[3]

In acknowledging the mystery of existence, the human mind seeks answers to fundamental questions of meaning and purpose in life. What is the purpose of the universe, and what is our role in it? Even if one answers that these questions are unanswerable, as an agnostic might, or if one answers that the universe is meaningless and our personal existence is a random cosmic fluctuation without significance, as some religions or atheists might, the fact remains that our mind is built to wrestle with these existential issues. Others, using the same capacities, seek God—as variously conceived as the answer to such questions. As James framed it: "Were one asked to characterize the life of religion in the broadest and most general terms possible, one might say that it consists of the belief that there is an unseen order, and that our supreme good lies in harmoniously adjusting ourselves thereto."[4]

Religious thought entails spiritual abstractions of the unseen, but they are, for the spiritually-minded, not at all unreal. The human capacity for symbolic thought flourishes here. The mind can attune, as James noted, to spiritual abstractions as comfortably as it does to the concrete concepts of the observable material world. As James explained

> This absolute determinability of our mind by abstractions is one of the cardinal facts in our human constitution. . . . We seek them, hold them, hate them, bless them, just as if they were so many concrete beings. And beings they are, beings as real in the realm which they inhabit as the changing things of sense are in the realm of space.[5]

DEALING WITH DEATH

The uniquely human ability to envision growing old or falling victim to an accident or disease is the sharp, painful edge of mental time travel. Death is not just an abstract concept understood at a distance—say, the way we understand the concept of the solar system. Rather, it is known through the deeply embodied emotional experience of losing others very close to us. It is known through the terrible downside of mental time travel, the capacity to imaginatively foresee our own death. According to terror management theory, awareness of our own mortality is potentially paralyzing and must somehow be mitigated.[6] The theory assumes that *Homo sapiens*, as with all other species, is designed for self-preservation, but it also assumes that we draw on symbolic thought and language capabilities to address the anxiety of that awareness. According to terror management theory:

> Cultural world views lend meaning through accounts of the origin of the universe, prescriptions for behavior, and explanations of what happens after death. Cultures differ radically in their specific beliefs, but share claims that the universe is meaningful and orderly, and that immortality is attainable, be it literally, through concepts of soul and afterlife, or symbolically, through enduring accomplishments and identifications (e.g., pyramids and novels, nations and causes, wealth and fame, ancestors and offspring). . . . A substantial proportion of human behavior is directed toward preserving faith

in a cultural world view and securing self-esteem in the service of death transcendence.[7]

Mortality can be made salient by asking someone to describe the emotions elicited by thinking about one's own death. Investigations using this technique have shown that "people's initial, conscious reaction to mortality salience is to deny their personal vulnerability to impending death (e.g., I am a healthy person, death is far off) and to suppress further death related thoughts."[8] In terms of the ensemble hypothesis developed here, the left-hemisphere interpreter first tries denial. However, once the immediate thought of death shifts out of the focus of attention in working memory, it can then linger and begin to affect ongoing cognitive processes. Specifically, the "heightened death-thought accessibility then triggers terror management defenses, which seem logically unrelated to the problem of death but bolster people's faith in their cultural world views and personal self-worth."[9] Once denial fails to manage the terror, the interpreter then defends the self's worldview, which includes political and religious beliefs.

Terror-management defenses of the sort described here have been observed in various cultures around the world. Importantly, it is not just religious beliefs that are bolstered by mortality salience. Attitudes toward politics, nationalism, prejudice, stereotyping, aggression, and social justice—in essence virtually any aspect of a person's worldview—are sensitive to terror-management defenses.[10]

The power of fear is unquestionable, and fear of death can indeed induce a state of terror. But is that all there is to religious impulses and theological inquiry? The studies supporting terror-management theory have shown that the mind copes with the threat of death by strengthening existing, already predominant beliefs about the world. That widely different kinds of beliefs respond to mortality salience, however, would seem to imply that terror management does not uniquely explain religious beliefs per se. Conservative or liberal political beliefs are held for a variety reasons, but they, too, are strengthened by the threat of death. Racial prejudice and nationalism also serve various psychological functions, but they also are strengthened by mortality salience. For example, nationalism is a way of losing self-identity in a massive cause shared by millions of fellow citizens. It serves a need to belong to a large social group independently of any role it may play in death transcendence. Similarly, spiritual inquiry serves the intellectual function of pondering the origin and purpose of the universe, and religion can bind people together as members of the same social group, just as nationalism can. Religion addresses living a purpose-filled life, and it is not just about death transcendence. Spirituality, then, can and does do more than simply serve as a defense mechanism against the terror of death. Stated differently, terror-management theory helps to account for why people defend their worldviews, but it is only a partial account of the religious mind.

DEALING WITH LIVING

Psychology in the past decade or so has become increasingly interested in the factors that contribute to human happiness and psychological adjustment. This field is known as positive psychology, and it can be contrasted with the historical tendency of the field to focus on maladjustment and abnormal mental and behavioral conditions. An interest in the links between religiosity and health is one avenue of such research.

Psychologists have examined the relationship between spirituality and well-being in a variety of ways. There is no shortage of psychological assessments of religiosity, with more than one hundred measures to choose from.[11] These include measurements of religious beliefs and practices, religious attitudes, religious values, religious development, religious commitment and involvement, spirituality and mysticism, and religious fundamentalism. Spirituality is sometimes conceptualized differently from religion, where the latter implies the outward practice of religion, such as attending church or endorsement of doctrinal beliefs. Spirituality refers more to inner experiences and attitudes that may or may not correspond with the doctrinal beliefs and practices of a given religion. So, it is possible to be highly spiritual without necessarily being religious. The distinction between intrinsic and extrinsic religiosity is similar. The utilitarian calculation of weighing benefits against costs is characteristic of extrinsic religiousness. Religion is seen as a means by which safety and social standing can be obtained. It is not the kind of religiosity that addresses the meaning of life and its purpose and direction. Intrinsic religiosity, on the other hand, provides an internal compass for navigating life that is relied upon for its own merits rather than for its utilitarian benefits.

Religion and spirituality have been associated with a variety of benefits in physical health. They are correlated with lower rates of heart disease, myocardial infarction, stroke, and high blood pressure.[12] Because coronary heart disease is such a pervasive health threat, the fact that religion and spirituality are connected to lower rates is of considerable interest from the standpoint of public health. Positive relationships have been broadly observed for men and women, for people of all ages, and for a variety of faiths, including Protestants, Catholics, Jews, Buddhists, and Muslims. Besides cardiovascular health, religiosity has also been associated with lower rates of cancer mortality.

Such physical health benefits are observed in those who experience religion in a healthy-minded way. However, it is also possible for religiosity to be harmful to health, as Kevin Seybold and Peter Hill described:

> The positive effects of religious and spiritual experience on health are based on the assumption that the experience itself is positive and healthy. Of course, religion and spirituality can also be pathological: authoritarian or blindly obedient, superficially literal, strictly extrinsic or self-beneficial, or conflict ridden and fragmented. Indeed such unhealthy religion or spirituality can have serious implications for physical health, having been associated with child abuse and neglect, intergroup conflict and violence, and false perceptions of control, with resulting medical neglect.[13]

Mental health is also related to religiosity. For example, religion and spirituality have been positively associated with general psychological functioning, well-being, and marital satisfaction, while correlating negatively with suicide and the abuse of alcohol and drugs.[14] Intrinsic religiosity has also been shown to reduce the dread of death. German researchers recruited participants in a Munich coffee shop soon after a terrorist attack in Istanbul and asked about their degree of fear that such attacks would take place in the near future on German soil.[15] They also completed a questionnaire that assessed intrinsic and extrinsic religious orientation.

The results showed that intrinsic religious belief precluded the need to invoke terror-management defenses. Individuals who scored low in intrinsic religiosity showed substantial evidence of strengthening their convictions about whether a terrorist attack was possible in Germany.[16] They actively sought information that confirmed their prior beliefs about their social world and violence. By contrast, those with strong religious beliefs turned to those beliefs to cope: they more often went to church, prayed that no more attacks would occur, and regarded their belief in God as an effective way to cope with the fear of terrorist attacks. Importantly, the researchers found that those with a high degree of intrinsic religiosity experienced a reduction in death-related thoughts, contrary to the usual consequence of being reminded of one's mortality. Terror-management defenses were unnecessary for this group because mortality salience "does not induce increased accessibility of death-related thoughts as it does for nonreligious or low intrinsically religious people."[17]

What might account for the positive relations between religiosity and health? There are many possible factors at work. For example, the positive emotion of gratitude could mitigate risks for depression and other mental health problems.[18] Gratitude involves recognizing a self-gain of value, while attributing that gain to the actions of someone else. It is thus a complex emotional response to receipt of a gift. For example, it is easy to imagine the feeling of gratitude that would come from an act of altruism, not reciprocal altruism, in which one anticipates that helping another person will result in that individual becoming obligated to return the favor, but a gift of help with no strings attached. There is also the sense of wonderment and awe provoked by the beauty of nature that can provoke gratitude to God, for those whose religious beliefs include a creator of the universe. Indeed, a deep perception that life itself is a gift from God evokes equally deep gratitude. Religious worship routinely incorporates expressions of praise and thanksgiving to God. The scriptural texts, prayers, rituals focus the attention of the worshiper on the gifts and mercies of God, inculcating a presence of gratitude. Perhaps positive emotional effects of such worship help in regulating the immune system, the autonomic nervous system, and the hypothalamus-pituitary-adrenal (HPA) axis.

Similarly, forgiveness may enter the equation, as "most religious accounts of optimal human functioning include the capacity to seek forgiveness and grant forgiveness as key elements of the well-functioning human personality."[19] The need to seek forgiveness depends on a social intelligence that understands how others think about one's transgressions. Being able to empathize with how another person must feel lies at the root of the experience of guilt. Of interest, in psychopaths these theory-of-mind capabilities are blunted, with the consequence that they experience neither empathy with their victims nor guilt about their transgressions. Ability to feel compassion and willingness to forgive the harm others have done requires not only an advanced social intelligence, but also the executive functions of working memory that resist the impulse to seek recompense or even revenge and suppress emotional responses of anger. Thus, forgiveness is another interesting dimension of spirituality that requires the ensemble of parts found only in the human mind. There may also be positive health consequences associated with forgiving others and seeking forgiveness by exercising the capacities of the mind's modern ensemble.

One intriguing lead is evidence that the anterior cingulate cortex (ACC) —a brain region serving the conflict-resolution function of executive attention in working memory—is affected by religious beliefs.[20] Environments that pose uncertainties, conflicts in goals that must be resolved, and errors in responding activate the ACC. In short, the situations that provoke anxiety and the need for control are signaled by the alarm bell of the ACC.

Researchers studied the ACC in a task called the *Stroop task*, in which participants must name the ink color of a series of color words. For half the words, the semantic meaning of the word (e.g., *blue*) matched the ink color and correct response (i.e., blue). For the other half, the correct response was incongruent with the word's meaning (e.g., *green* printed in blue ink). The incongruent trials pose a conflict between automatically reading the word and wanting to say green and inhibiting that immediate response and trying to say the correct ink color (i.e., blue). Participants slow down to resolve the conflict when making correct responses on incongruent trials, but sometimes they make errors and say green instead of blue.

Electroencephalographic recordings typically reveal strong neural activation of the ACC on incongruent trials. Yet people with strong religious convictions were able to minimize the alarm bell of the ACC in the Stroop task. Based on a standardized measure of religiosity, participants who scored high in religious zeal had less ACC activation to uncertainty and error in a simple laboratory decision task.[21] A second study found the same outcome for a milder form of religiousness; that is, the degree to which participants expressed a belief in God (ranging from certain that God exists to certain that God does not exist). Participants who were more certain about the existence of God exhibited less reactivity of the ACC and also greater accuracy in their decision making with fewer error responses. The researchers suggested that "conviction provides frameworks for understanding and acting within one's environment, thereby acting as a bulwark against anxiety-producing uncertainty and minimizing the experience of error."[22] Religious belief might, therefore, provide a means for coping with uncertainty and reducing anxiety and distress. The reduced ACC response could be part of a pathway that links to the HPA axis that mediates stress and immune responses.

Another possible health advantage of religious belief is as an aid to self-control of harmful behaviors.[23] Religious communities provide a social norm for behavior and encourage self-monitoring of behavior. Religious rituals also promote self-monitoring; for example, "preparing for weekly confession (for Catholics), the season of Lent (for many Christians), and the Yom Kippur holiday (for Jews) are supposed to involve examinations of one's spiritual and moral short-comings."[24] Besides aiding with monitoring of immoral or harmful behaviors, religious rituals, prayer, and meditation are all possible contributors to self-regulatory strength. Although a good deal more research is needed, the existing evidence is consistent with "the proposition that religion's ability to promote self-control or self-regulation can explain some of religion's associations with health, well-being, and social behavior."[25]

THE NEUROSCIENCE OF MYSTICAL EXPERIENCE

A striking variety of religious experience described by James is relatively rare but profound and highly distinctive. In mystical states of consciousness, James found four unique characteristics.[26] Mystical states of mind are ineffable, meaning that words cannot capture their qualities or meanings. Mystical states are at once both an expression of emotion and a state of knowing; they blur the difference between truth as an intellectual and rational state and awe or wonder as an intuitive and irrational commodity. Third, mystical states are relatively transient, lasting no more than thirty minutes or so. Fourth, and of considerable significance from the perspective of neuroscience and religion, the mystic feels carried away by a force far larger than the self. As James noted, "The mystic feels as if his own will were in abeyance, and indeed sometimes as if he were grasped and held by a superior power."[27] Although full-blown mystical experiences are rare, James recognized that many have experienced the rudiments of mysticism in a sudden appreciation or insight such as occurs in most people's everyday experience. Words said and deeds done many times in the past are suddenly comprehended with new meanings. For example, James cited the experience of Martin Luther in his sudden apprehension of the Nicene Creed:

> "When a fellow-monk," said Luther, "one day repeated the words of the Creed: 'I believe in the forgiveness of sins,' I saw the Scripture in an entirely new light; and straightaway I felt as if I were born anew. It was as if I had found the door of paradise thrown wide open."[28]

James further discussed the conversion vision of Saint Paul on the road to Damascus as well as the spiritual ecstasies of Saint Teresa of Ávila to illustrate the ineffable and otherworldly nature of mystical experiences: "The deliciousness of some of these states seem to be beyond anything known in ordinary consciousness."[29]

Neuropsychologists are naturally curious about what might be happening in the brain during mystical experiences of these sorts. One hypothesis is that they represent a seizure of a specific form in the temporal lobe of the brain. Patients with temporal-lobe epilepsy describe the hallucinations they experience during their seizures in mystical, spiritual terms, including "sensations of sudden ecstasy and religious awe; with increased interest in religion and even religious conversions; with out-of-body experiences; . . . and in some cases, with the perceived presence of God."[30] Another line of inquiry has drawn parallels between mysticism and the hallucinations and delusions of psychosis. Could it be that the religious visions of mystics are nothing more than temporal lobe seizures or psychotic breaks with reality?

James warned against the simplistic reduction of spiritual experiences to conditions of neurology. He called it the "the too simple-minded system" of medical materialism.[31] One flaw in such efforts to explain away mystic states as nothing more than odd neurology is in the incompleteness of such explanations. How can knowing the neurological states associated with a religious experience inform us about its spiritual significance and truth? Scientific or aesthetic or ethical thoughts can also be reduced in a simplistic way to neurological states, but the value or truth of these thoughts certainly are not decided one way or the other by knowing their neurology. James recognized that to single out only religious thoughts as inauthentic because of their neurological correlates is illogical and nothing more than prejudice against the notion that religious experience can be significant and true. James argued that

> scientific theories are organically conditioned just as much as religious emotions are; and if we only knew the facts intimately enough, we should doubtless see "the liver" determining the dicta of the sturdy atheist as decisively as it does those of the Methodist under conviction anxious about his soul. . . . To plead the organic causation of a religious state of mind, then, in refutation of its claim to possess superior spiritual value, is quite illogical and arbitrary, unless one has already worked out in advance some psychophysical theory connecting spiritual values in general with determinate sorts of physiological changes. Otherwise none of our thoughts and feelings, not even our scientific doctrines, not even our *dis*-beliefs, could retain any value of revelations of truth, for every one of them without exception flows from the state of its possessor's body at the time.[32]

That the brain is equipped for spiritual experiences could be just as readily seen as evidence for their reality. The normal assumption of evolutionary science is that the brain is adapted to pick up information from the environment that is important to the organism. Motion perception and color perception, for example, are seen as adaptive, and that is why the primate brain has these capacities. Could it be that, in the modern ensemble, the human brain is adapted to perceive reality at a spiritual level?

Another flaw in the reduction of mysticism to temporal-lobe epilepsy is the singularity of true mystical experience. An epileptic regularly experiences attacks; in serious cases, the temporal lobe seizes several times per week or even per day.[33] The hallucinations themselves are similarly repetitive, with the victim experiencing the same voice or the same intense feeling of knowing a hidden truth. Further, the hallucinations of epilepsy tend to be limited to a single modality—the victim sees a vision or hears a voice or feels a presence, but rarely all three. In contrast, "most mystics . . . experience only a handful of mystical encounters in a lifetime," and these entail a "high degree of sensory complexity" and "tend to be rich, coherent, and deeply dimensioned sensory experiences."[34] Psychotic breakdowns are also distinct in that the mystic returns to reality, shares the experience, functions normally, and in fact **is** "often seen as among the most respected and effective members of some societies."[35] This presents a sharp contrast to the highly distressed lives, often in social isolation, experienced by the chronically psychotic mental patient.

The ensemble hypothesis assumes that the modern human brain provides all the parts necessary for spiritual experiences and inquiries. The brain is able to think symbolically, to envision the future, and to infer hidden causes. Whether it is a true mystical state, a psychotic state, or a temporal-lobe seizure that causes the unusual state of consciousness, it is not surprising that all three might be understood in symbolic terms and interpreted as religious experiences. The aberrant electrical storms of temporal-lobe epilepsy and the delusions and hallucinations suffered in psychosis could plausibly be understood as spiritual experiences because the normal human brain is equipped for such experiences even in the absence of disease. In fact, there are a variety of religious practices, such as prayer and meditation, that can also lead to mystical states with no disease present at all.

Using methods of prayer and meditation specific to their religious traditions, practitioners can experience a feeling of merging into a union with the transcendent, where the self dissolves into something infinite and immortal. Islamic Sufis, Zen Buddhists, Tantric Buddhists, Kabbalistic Jews, Christian mystics, Hindus, Taoists, and Native American shamans have all reported such mystical experiences. To illustrate: "The goal of the contemplative rituals practiced by some Catholic mystics, for example, is to attain the state of *Unio Mystica*, the mysterious union that the mystic experiences as a sense of union with the actual presence of God. In Buddhism, the aim of meditative rituals is to encounter the ultimate oneness of everything by defeating the limiting sense of self generated by the ego."[36]

Andrew Newberg and Eugene d'Aquili captured the activity in the brain that transpires during these peak religious experiences using neuroimaging. Of special interest to them was the parietal lobe. In the posterior superior region of the parietal lobe, sensory signals from the senses of touch, vision, and hearing converge to create a mental representation of the body and its orientation in space. The left hemisphere in this parietal region computes a spatial representation of the self, whereas its analogue in the right hemisphere computes the surrounding physical space in which the self's body can reside.[37] Together, they allow the brain to maintain an understanding of where the self ends and the physical environment begins, and of how the body is spatially oriented from moment to moment.

In their studies, Franciscan nuns and Buddhist monks meditated and signaled the researcher when they achieved a peak experience. At this moment, a long-lasting radioactive tracer was injected into their blood stream. This allowed the use of single photon emission computed tomography (SPECT) in scanning their brain activity several minutes later so as not to interfere with the religious experience as it occurred during prayer or meditation. Compared with baseline activity in the parietal lobe, the SPECT scans of both the nuns and monks showed a decrease in activity in the posterior superior parietal region during the peak religious experience.[38] The region that would normally maintain a sense of how the self is oriented in space lowered its rate of neural firing during the peak religious experience. The neuroimaging findings correspond with the introspective reports of the mystical state of consciousness.

THE GENOME AND RELIGIOSITY

The present argument is that religion is a natural consequence of the design of the modern human brain. The advanced systems of working memory and social intelligence interact with mental time travel to enable spiritual inquiry and reasoning about theological questions, just as they enable moral reasoning. In fact, theological and moral questions are often intertwined. Of special importance is the role of the left-hemisphere interpreter that works with symbols and language, describing the abstractions of an unseen order.

Properties attributed to God—omniscience and omnipotence—can be understood and reasoned about perhaps only because of the words invented as labels for densely abstract concepts that are far from everyday percepts. They are like the mathematical concept of infinity; that is, difficult to really know in any way other than as a symbol. A primary function of the interpreter is to seek hidden causes for events perceived and thoughts held. It is thus essential in seeking answers to theological questions about the purpose and plan for the universe and humanity. From the current perspective, it is no surprise that human beings proceed both on faith as spiritual beings to understand theological problems and as scientific beings who can reason from empirical evidence. Both science and theology emerge from the ensemble of parts that constitute the brain of modern humans at least since the Upper Paleolithic.

Consistent with this view, a study with identical and fraternal twins has identified religious interests, attitudes, and values as being in part affected by genetic factors.[39] This makes sense if spirituality naturally emerges from the design of the modern human brain and genome. Investigators compared twins who were raised apart from each other with those raised together to try to assess the relative weight of the environmental and genetic influences. They took a variety of standardized measures of religious beliefs, interests, and activities. Their conclusion: about 50 percent of the individual variation in human populations with respect to religious interests is genetically influenced.[40] This finding runs strongly counter to the long-standing assumption that only cultural influence determines interest in religion. Yet it is one of very few studies that have examined religiosity in adults rather than in children, who are still under strong family influence.

Another study indicated that spiritual beliefs are already held in young children, suggesting they are in part a consequence of our genome and key brain networks. If family and culture indoctrinates children into believing in an afterlife, then one would expect that expressions of such a belief would grow with age. Kindergartners, late elementary school children, and adults were asked questions about what kinds of biological and mental states occur after death.[41] For example, does the brain still function, is there a need to eat, is it possible to love, and is it possible to want something or to know something? Compared with older children and adults, the younger children were more likely to attribute all of these abilities to a mind even after death. As the researcher noted, "This is precisely the opposite pattern that one would expect to find if the origins of such believes could be traced exclusively to cultural indoctrination. In fact, 'religious answers'—such as Heaven, God or spirits—among the youngest children were extremely rare."[42] Their intuitions about an afterlife were driven by nature, not nurture.

If it is correct that the ensemble of parts making up the modern human mind is responsible for the inevitable emergence of spiritual inquiry and religious belief, then it ought to be possible to find a link between these beliefs and the relevant brain networks mediating the parts. Research addressing such questions is just now beginning, but a study using functional magnetic resonance imaging (fMRI) in fact found such evidence.[43]

Religious beliefs were analyzed with the goal of understanding their underlying semantic relationships. Three dimensions were discovered that accounted for how people differ in their religious beliefs: God's perceived level of involvement in the world, God's perceived emotion, and a dimension that captured religious knowledge along a continuum of doctrinal versus experiential aspects. Doctrinal beliefs were expected to be associated with abstract semantics "whereas experiential knowledge engages networks involved in memory retrieval and imagery."[44] The fMRI results showed that the networks mediating theory of mind, abstract semantics, and the memory retrieval and visual imagery of mental time travel were critically involved. Such findings support the contention that it is our advanced social intelligence, mental time travel, and the capacity for abstract thoughts captured in symbols and language that underlie the human capacity for religion. Although social cognition, language, and logical reasoning have other important functions, as argued by the ensemble hypothesis, religious thought "likely emerged as a unique combination of these several evolutionarily important cognitive processes."[45]

A final argument favoring the notion that religion is endogenous to the structure of the human brain is its resiliency over time. Despite large changes in cultural climates, religion has stayed with us, and arguably always will, because it emanates from the ensemble of the modern brain. Beginning during the Enlightenment in the late eighteenth century and continuing through the nineteenth and twentieth centuries, intellectuals have predicted the end of belief in God and have done their best to bring it about. In their book *God Is Back*, John Micklethwait and Adrian Wooldridge write that

> Europe gave birth to succession of sages who explained, in compelling detail, why God was doomed. Karl Marx denounced religion as "the opiate of the masses." Émile Durkheim and Max Weber argued that the iron law of history was leading to "secularization." . . . Sigmund Freud dismissed religion as a neurosis that was designed to divert attention from man's real interest, sex. . . . It is hardly surprising that Marxist dictators such as Lenin and Mao tried to impose atheism by force."[46]

Even so, religion shows no signs of going away. If anything, the trends are in the direction of growth. For instance, "one estimate suggests that the proportion of people attached to the world's four largest religions—Christianity, Islam, Buddhism and Hinduism—rose from 67 percent in 1900 to 73 percent in 2005 and may reach 80 percent by 2050."[47] Even in Communist China, despite Mao's efforts, the "government's own figures show the number of Christians rising from fourteen million in 1997 to twenty-one million in 2006, with an estimated fifty-five thousand official Protestant churches and forty-six hundred Catholic churches." Including both house churches and the underground Catholic Church, "there are at least sixty-five million Protestants in China and twelve million Catholics—more believers than there are members of the Communist Party."[48]

Perhaps the strongest cultural headwind against the religious impulse of the human mind has come not from governments but from science. The conflict between science and religion is legendary and large. It is beyond the scope of this chapter to review the many battles that have unfolded since the time of Galileo, through Darwin, to the present. But the conflict was not and is not inevitable. Did God create the universe and human life or did evolution? This either/or choice misses the third possibility, that both statements are true. Perhaps evolution was the means by which God created the universe. Indeed, the noted Harvard botanist, Asa Gray, interpreted Darwin's *Origin of Species* in this manner in 1860.[49] Today, Francis Collins, director of the National Institute of Health in the United States and a leading scientist in the human genome project, sees in the discovery of DNA all the more reason to believe in God: "Evolution, as a mechanism, can be and must be true. But that says nothing about the nature of its author. For those who believe in God, there are reasons now to be more in awe, not less."[50] His view, however, is not shared by most members of the National Academy of Sciences, with only 10 percent professing a belief in God, with even fewer biologists (5 percent).[51]

If the ensemble of the modern human mind underlies both scientific and spiritual inquiry, then neither one is ever expected to triumph over the other. The eminent scientist Stephen Jay Gould offered a framework for understanding how this duality of human intellect can peacefully coexist in his book *Rock of Ages*. Science and religion constitute two separate teaching authorities or magisteria that address nonoverlapping sets of questions.[52] For example, what is our relationship, as human beings, with other living creatures? Darwin's theory of evolution is a teaching of science. It addresses whether living organisms are related in some manner through their genealogical links across geological time. For example, are apes similar to us in appearance and in some behaviors because we share a common ancestor with them? Does the greater degree of similarity between human DNA and chimpanzee DNA have some rational interpretation compared with the lesser degree of similarity between gorilla DNA or old world monkey DNA and human DNA? What does the DNA not coding genes and proteins do? Is it really just junk DNA? These kinds of questions, Gould concludes, fall within the magisterium of science, the "teaching authority dedicated to using the mental methods and observational techniques validated by success and experience as particularly suited for describing, and attempting to explain, the factual construction of nature."[53]

But now consider other kinds of questions about our relationships with other forms of life. Do we, if ever, have a right to drive another species to extinction through hunting or other means such as destroying habitat? Is the manipulation of the genetic code—transferring a gene from one species to another—a violation of the way the universe was intended to be? Is it, bluntly, immoral? Are the lives and feelings of the warm, fuzzy creatures we domesticate—our cats and our dogs—worth more than those of the animals we consume for food? Are the cattle, sheep, chickens, and pigs that we consume for food worth more than the insects we squash and swat, or, for that matter, the bacteria we routinely ignore until they cause illness? As Gould notes:

These questions address moral issues about the meaning of life, both in human form and more widely construed. Their fruitful discussion must proceed under a different magisterium, far older than science (at least as a formalized inquiry), and dedicated to a quest for consensus, or at least a clarification of assumptions and criteria, about ethical "ought," rather than a search for any factual "is" about the material construction of the natural world. This magisterium of ethical discussion and search for meaning includes several disciplines traditionally grouped under the humanities— much of philosophy, and part of literature and history, for example. But human societies have usually centered the discourse of this magisterium upon an institution called "religion."[54]

Gould's concept of nonoverlapping magisteria may help humanity to see scientific inquiry and spiritual inquiry as compatible rather than competitive. In physics, remarkable progress has been made in understanding the origins of the universe. Cosmologists have concluded that the universe originated in the "Big Bang" nearly fourteen billion years ago. Prior to that discovery it was possible to assume that the universe existed simply because it always had existed in a steady state. Acceptance of the Big Bang theory by cosmologists, however, inevitably raises the question: From what did the universe originate? This is *the* ancient question of theology and philosophy. What is the nature of the first cause or the prime mover of the universe?

The famous philosopher Anthony Flew abandoned atheism in part because of the discovery of the Big Bang and its implications for a creator. He concluded that an eternal God is as good an answer as any to the question of the origin of the universe. The famous physicist Stephen Hawking, on the other hand, saw no point in positing a creator. In his view the universe is self-contained. That is to say, the space-time continuum has neither a boundary nor an edge, meaning that is does not have a beginning or an end.[55] The expanding universe—a fact of the Big Bang—does not preclude a creator in Hawking's view, but he saw no need for one. As for the moment of the Big Bang, Hawking understands that $t = 0$, so one can in fact say that time itself had a beginning. But Hawking says that prior to this beginning, at $t < 0$, the concept of time is simply undefined and thus nothing can be said about it. Assuming this is correct, the question of first cause would seem to fall outside the magisterium of science. In Flew's words:

I concluded from this discussion that, even if it were agreed that the universe began with the big bang, physics must nonetheless remain radically agnostic: it is physically impossible to discover what, if anything, caused that big bang. . . . The moral of the story was that, ultimately, the issues at stake were philosophical rather than scientific.[56]

Our duality of intellectual inquiry exists only because the parts of the modern brain's ensemble make both possible. Coming to terms with this duality, then, will require learning to recognize the right teaching authority for the question at hand.

We, as twenty-first-century human beings, have inherited an amazing culture of spiritual exploration not only of the origins of the universe, but also of its purpose and destiny and our place within it. Religious thought and behavior has been part of human nature since our origins in prehistory. For the past four thousand years—just a fraction of our religious history—the monotheism of Judaism, Christianity, and Islam have informed the exploration of these mysteries for billions of people. The success of recent science—within the past one hundred years or so—should not precipitate amnesia for centuries upon centuries of religious insight. As Karen Armstrong reminded in her book *The History of Religion*:

My study of the history of religion has revealed that human beings are spiritual animals. Indeed there is a case for arguing that *Homo sapiens* is also *Homo religiosus*. Men and women started to worship gods as soon as they became recognizably human; they created religions at the same time as they created works of art. This was not simply because they wanted to propitiate powerful forces; these early faiths expressed wonder and mystery that seem always to have been an essential part of the human experience of this beautiful but terrifying world. Like art, religion has been an attempt to find meaning and value in life, despite the suffering that flesh is heir to.[57]

Chapter 11

TWENTY-FIRST-CENTURY MIND

By the time of the "human revolution" of the Upper Paleolithic, the human brain had the key design features that make the human mind so distinctive and of a different kind from any other species. The modern ensemble of mind included an advanced working memory and social intelligence. Both of these were likely prerequisites for the invention of spoken language and the social transmission of language from one generation to the next. The symbolic thought language built upon was not only a powerful tool for communicating with others, but also for communicating with ourselves as the voice of the mind within us. With an internal language for commenting and an ability to draw causal inferences, the left hemisphere of the human brain became specialized as the interpreter of consciousness. Its capacity for inferring hidden causes at work in physical and social events encountered in the world provided a powerful boost beyond the purely perceptual causation capabilities found in nonhuman primates. Intellectual inquiry using the ability to infer hidden causes was potently leveraged by the recollection of specific past events and the imagination of possible future events. Mental time travel, the interpreter, language, an understanding of the minds of others, and the executive functions of working memory together made for a formidable mind.

The origin of modern human beings, according to both archeological and genetic evidence, was in Africa. From there, human beings migrated to other parts of the world, eventually reaching all the continents of the earth. As Luigi and Francesca Cavalli-Sforza noted in *The Great Human Diasporas*, four key dates have emerged from archeological findings:

The first date refers to the oldest modern humans found in both Africa and the Middle East around one hundred thousand years ago. Available dates do not distinguish clearly which is older, but earlier skulls from Africa appear to show signs of a trend toward modern human forms, strengthening many archaeologists' conviction that *Homo sapiens sapiens'* birthplace was in Africa. The presence of modern human sites to the west and east of Suez one hundred thousand years ago suggests that the journey from Africa into Asia (or, less probably, vice versa) occurred at about that time. . . . The first human vestiges in Australia and New Guinea have been dated at fifty-five to sixty thousand years ago. . . . The other two dates are more recent and indicate the times of the occupation of Europe (probably from western Asia, around thirty-five to forty thousand years ago), and America. The date for the latter is still rather unclear, unfortunately, but it almost certainly fell between fifteen and thirty-five thousand years ago.[1]

Converging with the archeological findings, the degree to which different populations of the planet share genetic similarities and differences also shows the major split occurring first between Africans and non-Africans.[2] The settlement of southwestern Asia and Australia occurred next, according to archeological finds, and the genetic distance to African populations is correspondingly closest. Asia and Europe come next in measured genetic distance as well as in the archeological records, with northeastern Asia and America being the last regions of the globe settled by modern human beings.

In the Upper Paleolithic era, from around thirty-five to forty thousand years ago to about ten thousand years ago, modern human beings unquestionably occupied regions in Europe. They left behind unmistakable evidence of a mind like ours in their Paleolithic cave art and carved figurines.[3] An abundance of engraved bones and carved ivory show two- or three-dimensional representations of animals. The limestone caves at Altamira, Lascaux, Pech Merle, and dozens of other locations in southern Europe became the work sites of skilled drawers and painters. Included in their images are stenciled outlines of the human hand, such as those positioned next to the images of the famed spotted horses deep in the cave at Pech Merle. Thousands of miles from Pech Merle, in the vicinity of what is now Yellowstone National Park in Wyoming, stands Legend Rock, covered with petroglyphs. These carvings in the stone include a human handprint dated from about 10,700 years ago. Life-sized, the handprint was chipped out of the stone so it looked as if the hand had been pressed into the rock, as if the rock were soft clay.[4] Could there be a plainer symbol of the modern human mind's imprint on the world? The ensemble of the modern mind outlined in this book was there, thinking, talking, remembering, and creating at Legend Rock; by then the modern ensemble of mind had gone global.

The Great Leap Forward of the modern mind depended in part on an advanced working memory. One aspect of this was the enhancement of the executive-attention component of working memory, which allowed greater cognitive control and creative manipulation of mental representations. Another was the addition of a verbal working memory, which allowed the storage of words in the phonological form spoken by other human beings, as well as the storage of visual-spatial images of people, objects, and events. Whereas the latter enabled the power to learn, comprehend, and speak languages, the former enabled creative problem solving, planning, and mental flexibility.

Also in the creation of *Homo sapiens* came the capacity to think symbolically. Objects and events could be abstractly represented as words rather than as visual-spatial images. The modern mind was thus liberated from a purely perceptual train of thought. Besides bringing into working memory concrete objects and events that could be imaged like a picture, abstract concepts could also be represented as words and thought about through inner speech. The gift of language became a tool of solitary thought as well as a means of social communication. It is this capacity for symbolic thought—the use of words to refer to objects, events, thoughts, and concepts—that captures the essence of our species.

Language itself could not have been invented if it were not for an advanced social intelligence. We use language to communicate our thoughts and intentions to others and to try to influence their thoughts and intentions. This cognitive tool would not exist if it were not possible to understand other human beings as intentional agents with thoughts, desires, and perspectives of their own. We needed a theory of minds before we could invent a tool to change others' minds. An advanced social intelligence also enabled human beings to collaborate in ways beyond the game of language, to empathize with the feelings of others, and to transmit culture from one generation to the next through imitation and other means of social learning.

Regions of the left cerebral hemisphere, at least in the vast majority of human populations, were dedicated to the processing of language. Although the genetic basis for this adaptation is still not well understood, the human genome influences brain development in such a way that language inputs are preferentially processed by the left hemisphere, at least with respect to the sounds of words and their literal meanings. Two left-hemisphere networks known as Broca's area and Wernicke's area specialize, respectively, in the production and the comprehension of words. The left hemisphere also was dedicated to figuring out the hidden causes of events in the world. Inferring causation when it is not perceptually obvious is a powerful tool of analytic thought—science could not exist without it. Thus, in the left hemisphere of the human brain came a convergence that truly changed human history. Language—in an internalized form of inner speech—teamed up with causal inference to form the interpreter. The human mind could now explain the perceptions and thoughts that passed through conscious awareness. The inner chatter that accompanies most of our waking hours, and even accompanies some of our dreams, is of equal importance to human nature as the words we speak aloud to other human beings.

The ensemble of the human mind had one other component that possibly preceded some of the others. Namely, the system of long-term memory in humans underwent a specialization of declarative memory. Besides an ability to know a fact or a concept, human beings developed a capacity to recollect specific episodes from our past experiences. Besides knowing what occurred, we could reconstruct the when and where of it, too. Further, it allowed one to imagine an event occurring in the future using the same mental capacity. Human episodic memory, then, provided us with the capability of mental time travel, the ability to reverse the flow of time in our minds and to travel forward into an imaginary future. Because one of the functions of executive attention is to aid the effortful retrieval and reconstruction of past events, it seems likely that mental time travel required that an advanced system of working memory be in place first. It is also possible that language was a foundation for mental time travel. Certainly, human languages often provide a means for talking about the past versus the future. Yet did the invention of verb tense first require mental time travel or vice versa? It is impossible to say with certainty. We can be certain, however, that the modern ensemble includes a capacity to roam into the past of our autobiographical experience, to venture forward into the imaginary future, to report those episodes to other people, and to comment upon and explain the causal factors at work to ourselves.

The brain networks that enable the five parts of the modern ensemble were in place tens of thousands of years ago. Even so, the mind of a Cro-Magnon from that era in the Upper Paleolithic in what is now the south of France would have lacked the hundreds of centuries of cultural innovations that have shaped our minds. The avalanche of technological changes of the immediate past century alone has created a profoundly different cultural environment for the mind of today. Despite our biological identity, all seven billion plus of us today on earth would regard a Cro-Magnon person as "the other." Cro-Magnon prehistoric language, social customs, traditions, and beliefs would set them apart from any population now in existence on earth. Although Stone Age peoples—untouched by the modern world—were once still discoverable by anthropologists and psychologists, such isolation evaporated in the twentieth century. In an era when human beings have traveled to the moon and routinely look back on our planetary home from orbit, there no longer seems to be a place to hide apart from it all in prehistory. Have these cultural changes altered the way the contemporary mind functions compared with the mind of Cro-Magnon? If that is so, then how might the mind of the twenty-first century function differently as a result of the technological culture that now envelops us?

Such changes in mental functioning would appear inevitable because the mind is shaped by the culture as much as by the brain. The legendary neurophysiologist A. R. Luria conjectured how the human cerebral cortex acts as an organ of civilization. In the afterword to *Mind in Society*, Vera John-Steiner and Ellen Souberman described how Luria's views were shaped by his mentor Lev Vygotsky's emphasis on socialization in the formation of the human mind. In Luria's words:

> The fact that in the course of history man has developed new functions does not mean that each one relies on a new group of nerve cells and that new

"centers" of higher nervous functions appear. . . . The development of new "functional organs" occurs through the formation of *new functional systems*, which is a means for the unlimited development of cerebral activity. The human cerebral cortex, thanks to this principle, becomes an organ of civilization in which are hidden boundless possibilities, and does not require new morphological apparatuses every time history creates the need for a new function.[5]

The cerebral activity can undergo unlimited development by organizing new functional systems as a result of the social-learning experiences encountered during childhood. The culture in which the individual is raised can create new combinations of already-existing neural networks. The new functional systems can be built out of the existing modules and general resources of the brain, including working memory and long-term memory. For example, consider the three Rs of schooling. Reading, writing, and arithmetic are learned through immersion in a culture that uses, values, and teaches literacy and numeracy. The brain draws upon already existing neural centers for fine visual processing in order to perceive, represent, and manipulate letters and words in a written version of language. Similarly, written language can capitalize on the neural centers of Broca's area and Wernicke's area, which developed initially as brain adaptations for oral language. The processing of written language, then, is a new functional system built out of already-existing parts. Neither literacy nor numeracy requires the brain to change by evolving new adaptive structures across eons of geological time. Rather, within the existing structures of the cerebral cortex, new potential functions can find expression as human history advances and cultural evolution unfolds.

To bring our story to a close, this concluding chapter will examine how the modern ensemble might adapt to the remarkable technological innovations of the past century, indeed, the past decade. How might the Age of Information alter the mind? The Internet, laptops, smartphones, and 24/7 television now connect humanity in a global economy and information network that was unimaginable one hundred years ago. Such profound cultural innovation could precipitate the formation of new functional systems in the brain. Indeed, this has already happened before with the invention of writing and the cultural evolution of the literate mind.

THE MENTAL ORGAN OF WRITING

Prehistoric artists drew images and carved bone to represent objects and events witnessed in their environment. These drawings symbolically represented real, concrete things, but they did so in a literal or iconographic manner. The graphic symbol was directly related to the icon or image remembered in the mind of the prehistoric artist. Examples of such iconographic representations are abundant from tens of thousands of years ago, but only much more recently did human beings invent less literal symbols. Only much later did human beings begin to use symbols to refer to abstract ideas that could not be easily conveyed through pictures. Nearly thirty-five thousand years passed between the first appearance of representational art in bone engravings and ivory carvings and the invention of writing.

The origins of writing can be traced back to around 3200 BCE in Mesopotamia, in the region of modern day Iraq.[6] This writing system evolved gradually over a period of four thousand years from a system of tokens used by Neolithic farmers to keep track of food commodities, such as quantities of cereals. From these beginnings as a means of accounting, writing culturally evolved into an alternative mode of visible language and a powerful means of storing thought outside of the human brain. The innovation by the Greeks of the alphabetic writing we use today did not occur until around 750 BCE.[7] The Greek alphabet was phonetic, making it possible to represent spoken language in a precise way with a minimal load on memory. With a couple dozen symbols, all the possible morphemes could be represented. Reading could thus become learned more readily, and the foundation for literate cultures was thus laid.

Although it would take centuries for literacy to penetrate the vast majority of the world's population, the cultural innovation of writing had an unmistakable effect on human mental functioning. The invention of writing was perhaps inevitable, given the brain's capacity for symbolic thought and oral language, and the executive capacity for innovation in social domains. Writing, then, like morality and spirituality, emerged from the modern ensemble mind of humans.

The invention of writing and the spread of literacy fed back to alter brain functioning, the mind, and the societies in which we live in profound ways. As a consequence of cultural learning—through imitation and as a result of direct instruction—brain circuits are developed for decoding the symbols of written texts. Becoming a fluent reader entails training neural networks in the brain to recognize and comprehend the meaning of letters, words, phrases, sentences, and larger units of text structure. As with any visual stimulus, a word is first processed by the primary visual cortex in the occipital lobe at the rear of the brain. However, drawing on the existing specialization of the left hemisphere for language, a region just outside the primary visual cortex is activated by words in the left hemisphere—the fusiform gyrus, known as the word form area.[8] If presented with strings of letter that either are words (ANT) or could be words (GEEL), the word form area is more strongly activated than by letter strings that are nonsensical and not word like (TBBL). The brain seems to use an area that is commonly used in processing other kinds of visual stimuli that call for chunking together features and integrating them into a whole. Because the letters can be chunked into the whole of a word, this fusiform area of the left hemisphere becomes part of the mental organ for processing words as a person learns how to read.

The extent to which the brain builds on its existing networks for oral language to cope with the cultural innovation of writing is evident from the following fact. A child's ability to discriminate among the basic building blocks of oral language is an excellent predictor of the ease with which reading is learned.[9] Phoneme discrimination—a skill found in newborns because the brain is genetically prepared to tell one phoneme from another even in a foreign language not heard in the womb—is critical for later learning to the read the graphemes of written language. Between six and ten months, the sound distinctions that do not matter in the child's native language are lost as the phonological regions of the brain become attuned to the right phonemes for their oral language environment. There is a region in the left posterior part of the brain near the visual word form area that recodes the visual letters into the sounds of language. A second area of phonological processing is in the anterior region of the left hemisphere overlapping with Broca's area, which handles the articulation of sounds. This is the part of the brain involved in the phonological loop of verbal working memory and the inner voice of the interpreter.

As a third example of how reading uses the existing neural networks for oral language, a region just anterior to Broca's area in the left frontal cortex is activated when people pay attention to the meaning of a word.[10] Attending to the semantic interpretation of a word (e.g., hammer) occurs when, for example, one must think of a verb that describes a use for the object (e.g., pound). Neuroimages of the brain as people perform this usage test show activation in this frontal semantic area regardless of whether the word is heard or read.

The brain becomes an organ of civilization, as Luria called it, by adapting to the learning experiences of a literate upbringing. Studies of numeracy also conform to this principle. In a simple test of how the brain processes numbers, participants are asked to decide if a number is greater than or less than five. The numbers are presented visually either as Arabic digits (4) or as a spelled word (four). The spelled word activates the visual word form area in the left hemisphere because reading is involved in the task. Strikingly, the visual presentation of Arabic digits is processed very differently.[11] The activation is bilateral in the occipital and temporal lobes. The digits are a different kind of symbol from letters in that only digits draw upon the visual cortex as if they were pictures, making use of both hemispheres. The visual analysis is followed about forty milliseconds later by activation of both the left and right parietal lobes. It is here that the actual task of comparing the presented digit with the standard is carried out, followed by the decision that four or 4 is indeed less than five. The brain seems to represent a number line in the lower or inferior zone of the parietal lobe that allows an approximate judgment of magnitude.[12] This area of the brain is already specialized for the processing of spatial locations, so a number-line representation fits well with its more general purpose. When number comparisons are done using dot patterns rather than Arabic digits, brain-wave patterns look the same for both the perceptual and symbolic representations of numbers.

Prior to the invention of writing in any form, for tens of thousands of years, modern human beings thought and communicated entirely through spoken language. The thoughts of oral language take the form of a narrative or story. A narrative contains a beginning, a middle, and an end; it pits a protagonist against an antagonist; it follows a plot as the events of the story unfold. Such a structure is an apt way to convey foundational truths and beliefs about the world. For example, the fairy tales of Western culture are narratives about the basic issues of life: good versus evil, love versus hate, life versus death. All human cultures express their beliefs about the world through the medium of such shared communal stories. Consider the following:

> A gathering of modern postindustrial Westerners around the family table, exchanging anecdotes and accounts of recent events, does not look much different from a similar gathering in a Stone age setting. Talk flows freely, almost entirely in the narrative mode. Stories are told and disputed; and a collective version of recent events is gradually hammered out as the meal progresses.[13]

The interpreter of consciousness provides the fundamental cognitive machinery for storytelling. The inner voice explains the events in terms of their causes and consequences, and a narrative structure is imposed on the experience. It is this narrative structure that becomes the foundation of oral culture. The inner story is shared with others in the group, and such stories can be retold and passed from one generation to the next. Narrative thought, however, is distinctly different from the logical thought that human beings are also capable of employing. As Jerome Bruner observed, "A good story and a well-formed argument are different natural kinds . . . and the structure of a well-formed logical argument differs radically from a well-wrought story."[14]

Instead of thought proceeding in the form of a narrative, the alternative logical variety of thought proceeds from a premise to a conclusion or from specific data to a generalization. This kind of deductive and inductive thought is referred to as paradigmatic, analytic, or logicoscientific, and it owes much to the invention of writing.[15] A new mental organ arose from the innovations of civilization itself. Logicoscientific thought draws upon the external memory storage afforded by pictures, graphs, and writing. It depends upon the availability of knowledge stored outside the individual brain and mind in external symbolic stores. Working memory is not overwhelmed with simply retrieving the events of the past that form the grist of narrative thought. Instead, with the aid of external memory storage, it is free to think, to deduce conclusions and induce generalizations. Logical reasoning and analytic thought are different from the storytelling of oral culture, and they form the foundation of the theoretical culture of modern science. Again, it is not that narrative thought and oral culture have disappeared; rather, they have come to be accompanied by analytical thought and theoretical culture. Merlin Donald framed it this way: "In modern culture, narrative thought is dominant in the literary arts, while analytic thought predominates in science, law, and government."[16]

Consider: (1) formal arguments such as deductive reasoning from a syllogism proceeding from a major premise to a minor premise to a necessary conclusion; (2) systematic taxonomies that are common in the sciences; (3) operational definitions of theoretical concepts; and (4) formal methods of measurement. These are ways of thinking that require analysis rather than narrative. They function to produce logical and empirical truth rather than historical accounts or stories. In science, the end product is a theory that can integrate ideas and arguments. A successful theory does not only explain past events within its purview; it can also predict future events.[17]

In the contemporary practice of law, one can also see the strong dependence of analytic thought on the external memory afforded by written texts. Case law is argued by citing the legal reasoning of past cases. Constitutional law similarly requires reference to the reasoning laid out in previous decisions of the courts. It is difficult, if not impossible, to imagine how legal scholars could proceed were it not for the ability to store the facts and arguments of previous cases in external forms of memory. The writing of opinions—the advancement of legal scholarship—depends on reading the texts of the past. Literacy lies at its heart.

The external storage of pictures and written language permits one to reflect on facts and analyze their relationships in ways that the storytelling of oral language does not. Seeing the connections among ideas is aided by having them represented in a visual-spatial form in which they can be examined, moved around, and linked. This is difficult to do when ideas must be held solely in working memory as the transient sounds of oral language. An external representation of a problem permits one to contemplate and evaluate possible solutions more readily than when everything must be held mentally in one's head. Imagine, for example, playing chess blindfolded, such that with each move the position of one's pieces plus the opponents pieces must be held and updated in working memory. This would readily overload even a strong player's ability to entertain and evaluate alternative moves and anticipate the opponent's response. Only a master player, with an ability to retain the board's positions in well-learned chunks, is capable of playing without the external storage of a visible board. Thus, writing and analytical thought were symbiotic. With writing the modern human mind invented an external form of memory for the symbols of language. With it as a cognitive tool, "the human mind began to reflect upon the contents of its own representations, to modify and refine them."[18]

Importantly, the invention of writing and the development of analytical thought could not have occurred but for the ensemble of the modern human mind. Composing a written text requires planning ideas, expressing them in sentences, and reviewing ideas and the text already produced. The process of composing a text that modifies what one thinks about a topic requires advanced working memory, in particular, strong executive-attention skills. To use writing as a vehicle for thinking and expanding one's knowledge, one must simultaneously hold in mind what the author wants to say and an awareness of what the text actually says. There is a back and forth interaction between trying to say something in text and then seeing if it makes sense to the author. The author then can rethink the ideas and modify the text accordingly. By reviewing the text to ascertain whether it really reflects the author's thoughts, it becomes possible to think through the act of composing. Until a child acquires the maturation of the prefrontal cortex and develops sufficient executive attention, this juggling act of planning, sentence generating, and reviewing is not possible. Rather, text composition proceeds in the form of thinking about an idea, then putting it in a sentence, and then thinking of another idea. There is little to no interplay between the text as written and the writer's original ideas.[19]

Externalized memory further expands the mind by permitting collective cognition that relies on multiple individuals all sharing access to media held in common. A team of individuals can reason together and collaborate in problem solving when they all read and keep accessible a text that provides relevant data and facts. It is necessary only to refer to the location of information stored externally when communicating with each other rather than transmitting the information to each individual. Inferences can be drawn quickly, facts checked unequivocally, and common ground achieved with a minimal degree of effort given to oral communication. Collaborative cognition is certainly possible when individuals share the same perceptual space and are working to solve a concrete problem in the here and now (e.g., in a hunt). With the aid of externalized written and visuographic symbols, the benefits of collaboration are expanded immensely to abstract realms of thought far beyond the realms of the perceptual and motor.

The drawings, paintings, and sculptures of prehistory were the starting point for visuographic symbols. Although the emphasis has been placed on the transformational power of writing with a concise alphabetic code, numerous other forms also serve as external memory systems in theoretic culture. For example, films were introduced in the twentieth century and continue to dominate as a preferred means of storing stories, events, and factual knowledge externally in digital form, whether broadcast as television, downloaded from the Internet, or played as a video disk. History is filled to the brim with the symbolic notations that complement written language, however. Maps, astronomical charts, calendars, notations for dance choreography, musical scores, directorial play or film scripts, and engineering designs are all important examples. The expanse and richness of the theoretic cultures that have flourished in human history all depended on external memory storage. It is almost impossible to imagine human culture without the visuographic symbols that are taken so much for granted by us all.

Walter Ong, in *Orality and Literacy*, strongly advocated the view that writing, and later print, altered human thought processes. The technology of writing was necessary for the logical thinking that permeates history over the past several centuries. The Renaissance, which ended Europe's Middle Ages, and the Enlightenment of the seventeenth and eighteenth centuries paved the way for the scientific and technological societies of the contemporary world, but these historical developments would not have occurred without the invention of writing. As Ong phrased it, "Philosophy and all the sciences and 'arts' (analytic studies of procedures, such as Aristotle's *Art of Rhetoric*) depend for their existence on writing, which is to say they are produced not by the unaided human mind but by the mind making use of a technology that has been deeply interiorized, incorporated into the mental processes themselves."[20] Logicoscientific thought was facilitated over the centuries by the capacity to extend human memory through external symbol storage in the form of written texts.

In appreciating Ong's point, it is imperative to take a broad view of literacy that includes learning how to use reading and writing through schooling. As Donald Olson put it in *The World on Paper*, "Literacy in Western cultures is not just learning the abc's; it is learning to use the resources of writing for a culturally defined set of tasks."[21] Through literacy and the schooling that goes with it, for example, one can learn to use written texts, graphs, and diagrams in the conduct of scientific investigation. It is not enough to think only about the ability to produce or decode letters when thinking about how the acquisition of writing alters the mind. Literacy in this broad sense implies the awareness that written sources can be referenced for facts that one cannot recall; it implies an awareness that writing down one's thoughts can help to shape and articulate exactly what it is that one is trying to say. Literacy, properly understood, entails an awareness that the steps of a mathematical proof, the predictions of a scientific theory, or the entailments of a system of laws can only be fully grasped through their externalization as written notations. Just try to imagine doing mathematics, science, or law without writing.

Even so, as Walter Ong reminded us, the innovation of writing was also objectionable on the grounds that it could harm the skills of memory fostered by oral culture. In Plato's *Phaedrus*, Socrates argues that "writing destroys memory," for "those who use writing will become forgetful, relying on an external source for what they lack in internal sources."[22] Although writing may add to the mind, it may also subtract. External symbol storage expands our minds to a mode of logical, analytic thought and reduces reliance on internal memory at the same time. In oral cultures, knowledge must be passed down from generation to generation largely through narrative thought and spoken language. The only other means available are mimetic gestures and other nonverbal means of communication that likely preceded our capacity for oral language. Individual, internalized memory is thus essential for successful generational transmission of knowledge. Memory for oral language, therefore, is as fundamental to cultural evolution as genetic transcription and sexual reproduction are to biological evolution.

In his *Memory in Oral Traditions*, David Rubin reviewed the ways in which oral culture encouraged prodigious feats of individual internal memory. To illustrate, Avdo Medjedovi , a Slavic singer of epic poetry, was capable of recollecting poems as long as thirteen thousand lines that required sixteen hours to sing.[23] He knew a total of fifty-eight epics or about five hundred thousand lines—this stands in comparison to the twenty-seven thousand lines of the *Iliad* and *Odyssey* combined. Today, of course, outside of school, twenty-first-century human beings rarely memorize anything more than grocery lists and passwords to Internet accounts. Instead, we rely on the external stores of knowledge in written records, books, films, and the Internet.

Our reliance on writing as an external kind of memory has been shown in a laboratory experiment.[24] College students played a game of Concentration with a deck of thirty-six cards. One side of each card was a uniform, blue color, and on the other side was a picture (e.g., a flower, a sun, or one of several abstract designs). In the deck, each picture or design had one match. To begin Concentration, all the cards are laid down, blue side up. The first player selects a card and then turns over another card in search of its match. If successful, the player removes the pair and continues play. The turn ends when the player's choices result in a nonmatch. The goal is to find all matching pairs in as few turns as possible. Although the game is generally played by at least two players, it can be played as solitaire, as was done in the experiment. In one experimental condition, participants were allowed to write notes for themselves as they played the entire game. In a second experimental condition, participants again made notes as they wished, but in the seventh round their notes were taken away from them unexpectedly. Finally, in a control condition, no note taking was allowed and participants relied solely on internal memory. The results clearly showed an advantage for the writing group. By relying on external memory, they won the Concentration game in reliably fewer moves than the other two groups. The group interrupted in their use of notes performed almost as poorly as the group not allowed to take notes, indicating that they were relying on the external memory of their notes rather than memorizing the locations.

In the Concentration game, the player could focus more attention on remembering the identity of the cards because their locations could be stored externally. The external notes and internal memory worked together as part of a distributed cognitive system. More generally, in literate cultures, the information contained in books, magazines, maps, paintings, and film—to name just a few examples of written and other kinds of visuographic symbols—is symbiotic with internal human memory. Executive attention and the other components of working memory can be freed for a higher degree of reflective thought in such a symbiotic system. The analytic and logical thought of the literate mind is the result of freeing the mind from the narrative thought and memorization of oral culture.

THE INTERNET MIND

The way that twentieth-century science and technology transformed society is hard to grasp for anyone born in the past twenty-five years or so. Consider conditions one hundred years ago, when the automobile was just taking to the road in large numbers and competing with horses as a means of transportation. In the 1950s, when television first found its way into the homes of Americans, viewers were not infrequently treated to a test pattern occupying the screen when absolutely nothing was broadcast. With the 24/7 television of today, with its hundreds of channels, the days of three commercial networks and test patterns are as difficult to imagine as being on a wagon pulled westward by horses.

The influence of television, personal computers, cell phones, smartphones that combine the portability of a cell phone with the power of a computer, and the Internet are likely shaping the human mind into a new form, just as writing did. Unlike the introduction of writing in human culture, however, these information technologies have struck with lightning speed, on the scale of decades rather than centuries. Televisions, cell phones, and videocassette recorders (now already obsolete) saturated nearly 90 percent of US households within twenty years or so of their introduction.[25] Personal computer ownership and Internet access spiked equally rapidly.

Although writing was invented five thousand years ago, the vast majority of human beings remained illiterate until well into the most recent millennium. It was not until the development of moveable print in the fifteenth century that written texts were made available on a mass scale, and the schooling needed to learn to read inched forward only slowly even then. Although today the vast majority of the world's population is literate, it took thousands of years to get there. Thus, the manner in which writing altered the human mind was both subtle and glacially slow. Exposure to the Internet and global telecommunications, by contrast, is happening both on a massive scale and at breakneck speed. If the invention of writing created, in time, a new kind of functional mind based on literacy, then what mental organ will be forged by the telecommunications and computer revolution of our time? What will be the defining features of the twenty-first-century mind?

Color television was first introduced during President John F. Kennedy's years in the White House. But within three decades, by 1990, more than 90 percent of all US households had one or more to watch.[26] Although full-color audiovisual images had been part of the industrialized world since the 1920s, television brought the media into the home. Watching images on a screen was no longer a special event consigned to a movie theater. It was instead an everyday event that could be indulged for hours at a time, if one chose. In a survey of media in the home, the Annenberg Foundation reported in 2000 that children aged two to seventeen spent almost two and half hours viewing television each day.[27] Including video tapes, video games, computer use, and Internet browsing, the average amount of time spent in front of screens was over four and a half hours.[28]

The images of television, movies, and videogames are visually compelling to the human brain for several reasons.[29] First, the continuous motion of the action triggers the orienting component of attention. Unlike executive attention, orienting attention serves momentary perception rather than the sustained thought of working memory. Constant orientation to the environment directs attention to the external world of perception rather than the inward world of thought. Second, the brevity of events, interactions among people, and even the commercials in visual media enhance this focus on the external. Special effects heighten its perceptual allure. Scenes fade and dissolve into the next scene; montage combines disparate elements to guide the viewer through changes in place and time; music and lighting heighten emotional engagement; split-screen images, slow motion, and instant replay create a visual world unlike any encountered in real life. All this is well suited for portraying concrete actions that can be processed as visual-spatial images. In fact, it is so well suited as to be almost hypnotic in arresting and holding human attention within a fantasy world of images.

Whereas visual media give us images to enhance or even supplant the perceptual world of everyday experience, information technologies give us access to an essentially infinite number of electronic texts as well as images. The term Internet will be used in here as the proxy for all the related information technologies, such as personal computers and smartphones. The Internet links human beings with one another and with the images and text stored in external memory systems, and that is what matters most. Its superconnectivity is what affords the potential to shape the twenty-first-century mind. Even the division between television and the Internet is blurring: hypertext links within television programming make it more like surfing the Web, just as online video clips, movies, and other streams of imagery make our experience of the Internet more like television. The sheer quantity of texts and images available to human consciousness, and the ease with which they can be accessed, is revolutionary.

While providing massive external memory to augment the human brain, the associative links of the Internet use a similar associative structure to that found within our own internal semantic memory. The facts and concepts of semantic memory are linked together in a web of associations that allow one to jump from one idea to an entirely remote idea in just a few steps of associative thought. The Internet allows the same, but in a far deeper and richer database than any single human mind can contain. Besides providing access to an essentially infinite amount of information in the form of Web pages, the Internet offers the person-to-person connection of social networks. It offers a global virtual social world that is not constrained by the limitations of geographical proximity. By connecting people anywhere in the world, the Internet builds on the advanced social intelligence of human beings. For example, psychologists have asked whether mental and physical well-being can be enhanced through virtual social networks.[30]

TWO MINDS OF THE FUTURE

What effects do television and the Internet have on the functional capacities of the human mind? Taking for granted the power and pervasiveness of these technological innovations, how are they shaping the twenty-first century mind?

Nicholas Carr detailed the downside in his book *The Shallows*: the super-connectivity of images, texts, and people on the Internet is itself a problem because it distracts too much. To the extent that the Internet becomes the primary medium through which we now gather information about the world, the ability to navigate rapidly through a series of hyperlinks may weaken the mind's capacity for reflection on a few ideas of relevance and significance. Carr confesses that "what the Net seems to be doing is chipping away my capacity for concentration and contemplation."[31] With link after link, the Internet invites us into a continuous and clearly endless search through head-lines, abbreviated news reports, charts, posts to blogs, compelling images, video clips, podcasts, and so on. Just as the compelling nature of television draws our attention into its rapidly changing images, so, too, does browsing the Internet shape the mind so that it "wants and needs to take in and dole out information in short, disjointed, often overlapping bursts."[32]

The ubiquity of television in daily life—not just at home but in public spaces such as airports, dentists' offices, and restaurants—has given us a foretaste of how the twenty-first-century mind might be adapting to such technologies. Children ages one to three appear less focused in their play with toys and devote less time to them while a television is on in the background: "They begin to look like junior multitaskers, moving from toy to toy, forgetting what they were doing when they were interrupted by an interesting snippet of the show."[33]

Internet surfing and personal computer games are so compelling that some have worried that excessive exposure to them could even contribute to the development of Attention Deficit/Hyperactivity Disorder (ADHD). For example, one study reported an association between Internet addiction in elementary school children in Korea and symptoms of ADHD.[34] The exciting content of these media, relative to the mundane realities of life at home or in school, as well as the rapid sequence changes that repeatedly capture attention, could in theory spell trouble. The brain's networks of attention might develop differently in such a technologically laden environment compared with a purely natural one. A longitudinal study in New Zealand assessed this possibility by obtaining estimates from parents of how long their children watched television on weekdays at the ages of five, seven, nine, eleven, thirteen, and fifteen years.[35] The researchers then measured adolescent attention problems at the ages of thirteen and fifteen years based on self-report as well as reports from parents and teachers. The outcome showed that adolescents who watched two to three hours per day, and especially more than three hours per day, had more attention problems than those who watched less than two hours per day. The correlation was moderate in size, and the authors noted that children who already had attention problems often preferred to watch more television, indicating that the causal relation was actually bidirectional. In a similar study, efforts to measure attention problems and their link to exposure to a combination of television and video games found small to moderate, but statistically significant, correlations. For both boys and girls in middle childhood and in late adolescence, the total reported time engaged with screen media modestly predicted attention problems.[36] Because the findings were correlational, it is not possible to definitively conclude that excessive exposure to screen media causes attention problems. Still, the authors noted that "most of the research evidence thus far supports the conclusion that exposure to television and video games increases the risk for subsequent attention problems."[37]

Arguably, the revolution in applied computer science and telecommunications is formidable enough to distract the twenty-first-century mind. While this is of concern in children and adolescents still undergoing brain maturation, such findings in themselves may distract us from a larger truth. Could it be that we are lamenting the distraction of the Internet Age as Plato lamented the forgetfulness inflicted by the invention of writing? As became transparent centuries later, the literate mind brought us much more than forgetfulness. With its reliance on external memory storage, literacy also brought us a mode of logical-analytical thought that made possible the advent of modern science and the miracles of a technological world. Thus, distraction may be to the postliterate mind of the computer age what forgetfulness was to the literate mind of writing and external symbol storage. Forgetfulness aside, we would not have the scientific and technological culture of the twenty-first century had it not been for the widespread adoption of literacy. In an analogous way, the Internet, with its capacity for nearly instantaneous connection with virtually all knowledge and all people, could have a far more deep-seated effect on civilization than mere distraction.

E-mail, text messaging, and social networks allow the creation of a virtual social world in which one can actively participate. Because it is virtual, geographical distances no longer much matter. Language rather than physical proximity becomes the primary barrier to human connectivity. As long as two people share a common language, the virtual social world can encompass people from many countries and many cultures. Paradoxically, the borderless virtual space of the Internet seems to help shrink the world at the same time that it links together hundreds of millions of human beings. As the world's population now exceeds seven billion, the Internet can potentially create a global neighborhood through a common medium. Its audio and visual images, if not the languages, are shared across national and political borders. Assuming sufficient penetration of Internet access in a society, high and low income strata have access to the same free content, which opens doors to shared experiences, ideas, and conversations.

In addition to connections with people, the Internet provides connections with knowledge. As external memory for symbol storage, the Internet is massive and vastly superior in both capacity and access to nondigital libraries. Think of it as a library on steroids. The Internet provides access to essentially every known source of knowledge since the beginning of human history. Through the digitization of books, photographs, and other informational media, the Internet can integrate the written word with visual and auditory images far beyond the limits of printed books. The use of hypermedia also permits the retrieval of related knowledge in ways that books cannot match. The superconnectivity of the Internet may thus lay the groundwork for remarkable intellectual innovations. To see the Internet's potential, we must keep in mind that our existing scientific and theoretic culture draws chiefly on the power of external memory storage. The literate mind already overcame the limitations of retrieval from human long-term memory and the transient storage of working memory. Now, today's mind is poised to exploit an essentially unlimited external memory.

The expansion of external symbol storage in human history, from the famed library of ancient Alexandria to the Library of Congress or the British Library of the twentieth century, has been impressive. But the digital world of the Internet is transformative and still underappreciated. The digital storage of all printed materials—both texts and images—extends external memory storage to the limit of infinity while providing a manageable way of gaining access to any and all relevant information. The Internet is a "transbook . . . it is the book which can contain all books."[38]

The connections among people and stored knowledge afforded by the Internet can be the catalyst for a collective form of intelligence. The scientific and theoretic culture made possible by literacy is just the beginning. By pooling and cross-germinating our ideas, a connected contemporary mind can culturally evolve from the brain of our Paleolithic ancestors. Hidden within the cerebral cortex of the human brain lies a new organ of civilization, one that could be expressed by networking ideas and people massively and cross-culturally. To see this point, keep in mind that cultural innovations of the past have always arisen out of human connectivity. Innovation depends on the collision of ideas fostered by human interactions.[39]

For example, the rate of cultural change in human populations has always depended on the degree to which individuals engage in trade in the economic sphere. Being part of a trade route in the ancient world meant everything. Istanbul, for example, prospered because it was at the center of Asia to the east and Europe to the west. Cultural innovations and the changes in our functional mental capacities that came with them have always depended on human interactions. Cities, both in the ancient and in the modern world, have been and continue to be the centers of technological innovation. It is not a surprise that Silicon Valley in the San Francisco Bay area was the birthplace of the revolution in personal computing and telecommunications rather than, say, Death Valley. The dense population of the Bay area, with its universities, research companies, and sources of venture capital in close proximity, facilitated technological innovation. Dense populations allow human minds to meet and cross-fertilize much more readily than sparse populations.

The collective intelligence that drives innovations in technology requires collaborative efforts. The capability of the Internet to link ideas and people eliminates the problem of geography. As long as the texts and images are mutually understood, collaboration is now possible regardless of where people live. If dense cities and trade routes were the incubators of innovation in the past, the Internet opens a new door. Perhaps face-to-face contact—the key advantage of being in the right city at the right time—will no longer matter. In short, the Internet provides a new kind of trade route of ideas and people through a virtual landscape. Collaborative innovation on a worldwide scale is thus only now possible for the first time in human history.

Don Tapscott and Anthony Williams developed this theme in the economic realm in their book *Wikinomics: How Mass Collaboration Changes Everything*. One illustration of successful global collaboration enabled by the Internet is the case of a gold-mining company that released all its proprietary geological data—going back five decades—and challenged the global community to find new deposits. The "Goldcorp Challenge" offered prize money to entice submissions from geologists, graduate students, and other participants from diverse scientific backgrounds who identified more than fifty new target areas unexplored by the company in the past—80 percent of these struck gold, and since the challenge was initiated, an astounding eight million ounces of gold have been found."[40]

In the scientific realm, Michael Neilsen makes the case in *Reinventing Discovery* that "to historians looking back a hundred years from now, there will be two eras of science: pre-network science, and networked science."[41] The Internet could dramatically increase the rate of progress in science and in applying new discoveries to solving global problems. To illustrate, a multiplayer online game was recently devised to solve problems in molecular biology.[42] The three-dimensional structures of proteins are specified by the sequence of their amino acids, but the number of possible ways in which they can be structured exceeds our computational power. Microbiologists have had little success in predicting protein structure with software because the solution space is too large for all but the smallest of proteins. Collective human intelligence was brought to bear on the problem by creating an online video game called Foldit. A puzzle protein is posted online for a fixed interval of time and players from around the world interactively reshape the protein with a goal of minimizing its energy profile. Players may compete as a soloist or work in groups. Competition is encouraged by posting each soloist or group score from best to worst. The top-ranked Foldit players excelled in finding correct solutions, and players working in collaboration discovered new search strategies not examined by existing software. Foldit demonstrates the potential of tapping collaborative problem solving on a massive scale through the Internet.

The Internet capitalizes on all parts of the modern ensemble of mind. Creative problem solving enabled by the executive functions of working memory teams up with the symbolic thought of written language. Our advanced social intelligence may find its fullest expression through the massive, global connectivity of the Internet. Mental time travel can now draw not only upon the episodic memory storage of the brain, but also upon a vast external store of history. The immediate accessibility of all recorded knowledge as text and images—a transbook that contains all books—provides an unprecedented resource for the imagination. Our human capacity to envision the future has never had a richer source of possibilities. The superconnectivity of the Internet may thus be a technological invention capable of nurturing social networks, opening knowledge access to all, and encouraging collaborative problem solving on a global scale. If the Internet's promise is actualized, then the worry about technology-induced distraction will perhaps subside. A distracted mind may be the price we pay for a connected mind, much as forgetting was the price paid for the advances writing brought us. If speaking was crawling, and writing was walking, then we are now running with the Internet. Where, then, is our destination?

NOTES

CHAPTER 1. ORIGINS

1. Randall White, *Prehistoric Art: The Symbolic Journey of Humankind* (New York: Harry N. Abrams, 2003), p. 97.

2. Melanie Proust et al. "Genotypes of Predomestic Horses Match Phenotypes Painted in Paleolithic Works of Cave Art," *Proceedings of the National Academy of Sciences* 108, no. 46 (2011): 18626.

3. Margaret W. Conkey, "A History of the Interpretation of European 'Paleolithic Art': Magic, Mythogram, and Metaphors for Modernity." In *Handbook of Human Symbolic Evolution*, eds. Andrew Lock and Charles R. Peters (Oxford: Clarendon Press, 1996), pp. 288–95.

4. Richard G. Klein and Blake Edgar, *The Dawn of Human Culture* (New York: Nevraumont Publishing Company, 2002), p. 261.

5. Sally McBrearty and Alison S. Brooks, "The Revolution That Wasn't: A New Interpretation of the Origin of Modern Human Behavior," *Journal of Human Evolution* 39 (2000): 456.

6. Stephen J. Gould, *The Structure of Evolutionary Theory* (Cambridge, MA: Belknap Press of Harvard University Press, 2002), p. 914.

7. Ibid., pp. 765–69.

8. Gregory Cochran and Henry Harpending, *The 10,000 Year Explosion: How Civilization Accelerated Human Evolution* (New York: Basic Books, 2009), p. 77.

9. Steve Olson, *Mapping Human History: Genes, Race, and our Common Origins* (Boston: Houghton Mifflin Company, 2002), p. 16.

10. Ibid., p. 17.

11. Wen-Hsiung Li and Matthew A. Saunders. "The Chimpanzee and Us," *Nature* 437 (September 1, 2005): 50.

12. Carina Dennis, "Primate Evolution: Branching Out," *Nature* 437 (September 1, 2005): 17–19.

13. C. Owen Lovejoy, "Reexamining Human Origins in Light of *Ardipithecus ramidus*," *Science* 326 (October 2, 2009): 74e1, doi:10.1126/science.1175834.

14. Peter H. Raven and George B. Johnson, *Biology*, 5th ed. (Boston: McGraw-Hill, 1999), p. 458.

15. Olson, *Mapping Human History: Genes, Race, and our Common Origins*, p. 3.

16. Roger Lewin, "Mitochondrial Eve: The Biochemical Route to Human Origins," *Mosaic* 22, no. 3 (1991): 48–49.

17. Raven and Johnson, *Biology*, p. 459.

18. Richard Thompson, *The Brain: A Neuroscience Primer*, 3rd ed. (New York: Worth Publishers, 2000), p. 3.

19. Jackson Beatty, *The Human Brain: Essentials of Behavioral Neuroscience* (Thousand Oaks, CA: Sage Publications), p. 52.

20. Sean B. Carroll, "Genetics and the Making of *Homo sapiens*," *Nature* 422 (2003): 849.

21. Richard G. Klein, *The Human Career: Human Biological and Cultural Origins*, 2nd ed. (Chicago: University of Chicago Press, 1999), p. 580.

22. Ralph Holloway, "Evolution of the Human Brain." In *Handbook of Human Symbolic Evolution*, eds. Andrew Lock and Charles R. Peters (Oxford: Clarendon Press, 1996), p. 80.

23. Robert Sean Hill and Christopher W. Walsh, "Molecular Insight into Human Brain Evolution," *Nature* 437 (September 1, 2005): 64.

24. Carroll, "Genetics and the Making of *Homo sapiens*," p. 851.

25. Chris P. Ponting and Gerton Lunter, "Human Brain Gene Wins Genome Race," *Nature* 443 (September 2006): 149.

26. Thompson, *The Brain*, p. 8.

27. Ibid, p. 14.

28. Ibid, p. 15.

29. Ibid, p. 17.

30. Beatty, *The Human Brain*, pp. 59–60.

31. Paul D. MacLean, "On the Evolution of the Three Mentalities of the Brain." In *Origins of Human Aggression: Dynamics and Etiology* (New York: Human Sciences Press, 1987), pp. 29–38.

32. Ibid., p. 39.

33. Joseph E. LeDoux, "Emotion Circuits in the Brain," *Annual Review of Neuroscience* 23 (2000): 157.

34. Charles Darwin, *The Descent of Man and Selection in Relation to Sex*, 2nd ed. (New York: Collier, 1905), p. 170.

35. Ibid., pp. 170–71.

CHAPTER 2. EXECUTIVE WORKING MEMORY

1. P. S. Goldman-Rakic, "Cellular Basis of Working Memory," *Neuron* 14 (1995): 483.

2. Thomas Wynn and Frederick L. Coolidge, "A Stone-Age Meeting of Minds," *American Scientist* 96 (January–February 2008): 46.

3. Steven Mithen, *The Prehistory of the Mind: The Cognitive Origins of Art, Religion, and Science* (London: Thames and Hudson, 1996), p. 155.

4. Wynn and Coolidge, "A Stone-Age Meeting of Minds," p. 49.

5. Ibid.

6. Edward E. Smith and John Jonides, "Working Memory: A View from Neuroimaging," *Cognitive Psychology* 33 (1997): 11.

7. Morris Moscovitch, Gordon Winocur, and Marlene Behrmann, "What Is Special about Face Recognition? Nineteen Experiments on a Person with Visual Object Agnosia and Dyslexia but Normal Face Recognition," *Journal of Cognitive Neuroscience* 9 (1997): 587–89.

8. Jennifer Steeves et al., "Abnormal Face Identity Coding in the Middle Fusiform Gyrus of Two Brain-Damaged Prosopagnosic Patients," *Neuropsychologia* 47 (2009): 2584.

9. David C. Geary, *The Origin of Mind: Evolution of Brain, Cognition, and General Intelligence* (Washington, DC: American Psychological Association, 2005), pp. 128–29.

10. Mithen, *The Prehistory of the Mind*, pp. 163–64.

11. Wynn and Coolidge, "A Stone-Age Meeting of Minds," pp. 45–46.

12. Michael S. Gazzaniga, Richard B. Ivry, and George R. Mangun, *Cognitive Neuroscience: The Biology of the Mind* (New York: W. W. Norton & Company, 1998), p. 426.

13. Ibid., p. 425.

14. K. Semendeferi et al., "Humans and Great Apes Share a Large Frontal Cortex," *Nature Neuroscience* 5 (March 2002): 273.

15. Ibid., p. 274.

16. Nelson Cowan, "The Magical Number 4 in Short-Term Memory: A Reconsideration of Mental Storage," *Behavioral and Brain Sciences* 24 (2001): 87.

17. Joël Fagot and Carlo De Lillo, "A Comparative Study of Working Memory: Immediate Serial Recall in Baboons (*Papio papio*) and Humans," *Neuropsychologia* 49 (2011): 3872.

18. Goldman-Rakic, "Cellular Basis of Working Memory," p. 483.

19. Semendeferi et al., "Humans and Great Apes," p. 272.

20. Alan Baddeley, Susan Gathercole, and Costanza Papagno, "The Phonological Loop as a Language Learning Device," *Psychological Review* 105 (1998): 158.

21. Susan E. Gathercole and Alan D. Baddeley, *Working Memory and Language* (Hillsdale, NJ: Lawrence Erlbaum, 1993), p. 41.

22. Alan Baddeley et al., "The Phonological Loop," p. 159.

23. Ibid., pp. 161–62.

24. Smith and Jonides, "Working Memory," p. 12.

25. Semendeferi et al., "Humans and Great Apes," p. 275.

26. Peter R. Juttenlocher and Arun S. Dabholkar, "Regional Differences in Synaptogenesis in Human Cerebral Cortex," *Journal of Comparative Neurology* 387 (1997): 167.

27. Elizabeth R. Sowell, Paul M. Thompson, Colin J. Holmes, Terry L. Jernigan, and Arthur W. Toga, "In Vivo Evidence for Post-Adolescent Brain Maturation in Frontal and Striatal Regions," *Nature Neuroscience* 2 (1999): 859–60.

28. Ibid., p. 860.

29. Semendeferi et al., "Humans and Great Apes," p. 272.

30. Michael I. Posner and Mary K. Rothbart, *Educating the Human Brain* (Washington, DC: American Psychological Association, 2007), p. 60.

31. Ibid., pp. 82–86.

32. Ibid.

33. Akira Miyake and Naomi P. Friedman, "The Nature and Organization of Individual Differences in Executive Functions: Four General Conclusions," *Current Directions in Psychological Science* 21 (2012): 8.

34. Ibid., p. 11.

35. Posner and Rothbart, *Educating the Human Brain*, p. 91.

36. Ibid., p. 92.

37. Walter Mischel, Yuichi Shoda and Phillip K. Peake, "The Nature of Adolescent Competencies Predicted by Preschool Delay of Gratification," *Journal of Personality and Social Psychology* 54 (1988): 688–91.

38. Randall W. Engle et al., "Working Memory, Short-Term Memory, and General Fluid Intelligence: A Latent Variable Approach," *Journal of Experimental Psychology: General* 12 (1999): 324.

39. John Duncan et al., "A Neural Basis for General Intelligence," *Science* 289 (July 2000): 459.

40. Posner and Rothbart, *Educating the Human Brain*, p. 111.

41. Simon M. Reader and Kevin N. Laland, "Social Intelligence, Innovation, and Enhanced Brain Size in Primates," *Proceedings of the National Academy of Sciences* 99 (April 2002): 4437–38.

CHAPTER 3. SOCIAL INTELLIGENCE

1. Richard G. Klein and Blake Edgar, *The Dawn of Human Culture* (New York: Nevraumont Publishing Company, 2002), p. 261.

2. Randall White, "On the Evolution of Human Socio-Cultural Patterns." In *Handbook of Human Symbolic Evolution*, eds. Andrew Lock and Charles R. Peters (Oxford: Clarendon Press, 1996), pp. 246–51.

3. Xiaiohong Wu et al., "Early Pottery at 20,000 Years Ago in Xianrendong Cave, China," *Science* 336 (June 2012): 1696.

4. Ibid., p. 1699.

5. W. Michael Cox and Richard Alm, "You Are What You Spend," *New York Times*, Sunday Opinion, February 10, 2008, final edition, p. 14.

6. Michael Tomasello, *The Cultural Origins of Human Cognition* (Cambridge, MA: Harvard University Press, 1999), p. 2.

7. Ibid., p. 38.

8. Ibid., p. 37.

9. Darrin R. Lehman, Chi-yie Chiu, and Mark Schaller, "Psychology and Culture," *Annual Review of Psychology* 55 (2004): 697–700.

10. Ibid., p. 698.

11. Ibid.

12. Seth J. Schwartz, Jennifer B. Unger, Byron L. Zamboanga, and José Szapocznik, "Rethinking the Concept of Acculturation: Implications for Theory and Research," *American Psychologist* 65 (2010): 242–43.

13. Andrew N. Meltzoff, Patricia K. Kuhl, Javier Movellan, and Terrence J. Sejnowski, "Foundations for a New Science of Learning," *Science* 325 (July 2009): 284.

14. Peter Mundy and Lisa Newell, "Attention, Joint Attention, and Social Cognition," *Current Directions in Psychological Science* 16 (2007): 270.

15. Rechele Brooks and Andrew N. Meltzoff, "The Development of Gaze Following and Its Relation to Language," *Developmental Science* 8 (2005): 535.

16. Mundy and Newell, "Attention, Joint Attention, and Social Cognition," p. 271.

17. Tomasello, *The Cultural Origins of Human Cognition*, p. 63.

18. Ibid., p. 68.

19. Simon Baron-Cohen, Alan M. Leslie, Uta Frith, "Does the Autistic Child Have a 'Theory of Mind,'" *Cognition* 21 (1985): 39.

20. Ibid., pp. 39–42.

21. Ibid., pp. 37–39.

22. Tomasello, *The Cultural Origins of Human Cognition*, p. 81.

23. Meltzoff et al., "Foundations for a New Science of Learning," p. 285.

24. Ibid.

25. Andrew Whiten, "The Second Inheritance System of Chimpanzees and Humans," *Nature* 437 (September 2005): 52–53.

26. Ibid, p. 54.

27. Tomasello, *The Cultural Origins of Human Cognition*, p. 29.

28. Ibid., pp. 29–30.

29. Ibid., p. 30.

30. Whitten, "The Second Inheritance System," p. 54.

31. Ibid.

32. Tomasello, *The Cultural Origins of Human Cognition*, p. 30.

33. Ibid., p. 35.

34. Merlin Donald, *Origins of the Modern Mind* (Cambridge, MA: Harvard University Press, 1991), p. 149.

35. Brian Hare, "From Nonhuman to Human Mind: What Changed and Why?" *Current Directions in Psychological Science* 16 (2007): 61.

CHAPTER 4. LANGUAGE

1. Charles Darwin, *The Descent of Man and Selection in Relation to Sex*, 2nd ed. (New York: Collier, 1905), p. 171.

2. André Parrot, *The Tower of Babel* (New York: Philosophical Library, 1955), p. 17.

3. Ibid., pp. 18–23.

4. Jean Aitchison, *The Seeds of Speech: Language Origin and Evolution* (Cambridge: Cambridge University Press), p. 5.

5. Joel Davis, *Mother Tongue: How Humans Create Language* (New York: Birch Lane Press, 1994), p. 26.

6. L. L. Cavalli-Sforza and F. Cavalli-Sforza, *The Great Human Diasporas: The History of Diversity and Evolution* (Reading, MA: Addison-Wesley Publishing Company, 1995), pp. 169–70.

7. Ibid., pp. 180–82.

8. Merritt Ruhlen, *The Origin of Language: Tracing the Evolution of the Mother Tongue* (New York: John Wiley and Sons, 1994), p. 163–64.

9. Luigi Luca Cavalli-Sforza, Alberto Piazza, Paolo Menozzi, and Joanna Mountain, "Reconstruction of Human Evolution: Bringing Together Genetic, Archaeological, and Linguistic Data," *Proceedings of the National Academy of Sciences* 85 (1988): 6005.

10. Ibid.

11. Ruhlen, *The Origin of Language*, pp. 27–28.

12. Terrence W. Deacon, *The Symbolic Species: The Co-Evolution of Language and the Brain* (New York: W. W. Norton & Company, 1997), p. 22.

13. Ruhlen, *The Origin of Language*, p. 31.

14. Ibid.

15. Ibid., p. 30.

16. Ronald T. Kellogg, *Cognitive Psychology*, 2nd ed. (Thousand Oaks, CA: Sage Publications, 1995), p. 269.

17. Ibid., p. 308.

18. Russell A. Poldrack and Anthony D. Wagner, "What Can Neuroimaging Tell Us about the Mind: Insights from Prefrontal Cortex," *Current Directions in Psychological Science* 13 (2004): 177–78.

19. Kellogg, *Cognitive Psychology*, pp. 302–304.

20. Ibid., pp. 281–83.

21. Ibid., pp. 272–73.

22. Herbert Clark, *Using Language* (Cambridge: Cambridge University Press, 1996), p. 3.

23. Klaus Zuberbühler, "The Phylogenetic Roots of Language: Evidence from Primate Communication and Cognition," *Current Directions in Psychological Science* 14 (2005): 127.

24. Ibid., p. 128.

25. Morton H. Christiansen and Simon Kirby, "Language Evolution: Consensus and Controversies," *TRENDS in Cognitive Sciences* 7 (July 2003): 301.

26. Derek Bickerton, *Language and Species* (Chicago: University of Chicago Press, 1990), p. 14.

27. Roger Fouts, *Next of Kin: What Chimpanzees Have Taught Me about Who We Are* (New York: William Morrow and Company, 1998), p. 24.

28. Phillip Liberman, *Eve Spoke: Human Language and Human Evolution* (New York: W. W. Norton & Company, 1998), pp. 45–61.

29. Cecilia S. L. Lai et al., "FOXP2 Expression during Brain Development Coincides with Adult Sites of Pathology in Severe Speech and Language Disorder," *Brain* 126 (2003): 2455–62.

30. Wolfgang Enard et al., "Molecular Evolution of FOXP2, a Gene Involved in Speech and Language," *Nature* 418 (August 2002): 871.

31. Beatrice T. Gardner and R. Allen Gardner, "Evidence for Sentence Constituents in the Early Utterances of Child and Chimpanzee," *Journal of Experimental Psychology: General* 104 (1975): 244–48.

32. E. Sue Savage-Rumbaugh and Duane M. Rumbaugh, "The Emergence of Language." In *Tools, Language and Cognition in Human Evolution*, eds. Kathleen R. Gibson and Tim Ingold (Cambridge: Cambridge University Press), p. 90.

33. Ibid., pp. 92–99.

34. Bickerton, *Language and Species*, pp. 107–109.

35. Kellogg, *Cognitive Psychology*, pp. 283–88.

36. Merlin Donald, *Origins of the Modern Mind: Three Stages in the Evolution of Culture and Cognition* (Cambridge, MA: Harvard University Press), p. 211.

37. Ibid., pp. 165–68.

38. Susan Goldin-Meadow "Talking and Thinking with Our Hands," *Current Directions in Psychological Science* 15 (2006): 38.

39. Ibid., p. 34.

CHAPTER 5. THE INTERPRETER OF CONSCIOUSNESS

1. Michael S. Gazzaniga, Richard B. Ivry, and George R. Mangun, *Cognitive Neuroscience: The Biology of Mind* (New York: W. W. Norton & Company, 1998), pp. 330–31.

2. Ibid., pp. 344–45.

3. Michael S. Gazzaniga, *The Mind's Past* (Berkeley: University of California Press, 1998), pp. 1–2.

4. Ibid., p. 133.

5. Michael S. Gazzaniga, "Cerebral Specialization and Interhemispheric Communication: Does the Corpus Callosum Enable the Human Condition?" *Brain* 123 (2000): 1316.

6. Ibid., p. 1318.

7. Ibid., p. 1316.

8. George Wolford, Michael B. Miller, and Michael Gazzaniga, "The Left Hemisphere's Role in Hypothesis Formation," *Journal of Neuroscience* 20 (2000 RC64): 1–2.

9. Ibid.

10. Gazzaniga, *The Mind's Past*, p. 134.

11. Ibid., p. 136.

12. Ibid.

13. Jennifer A. Whitson and Adam D. Galinsky, "Lacking Control Increases Illusory Pattern Perception," *Science* 322 (October 2008): 115.

14. Ibid.

15. Ibid., p. 116.

16. Matthew E. Roser et al., "Dissociating Processes Supporting Causal Perception and Causal Inference in the Brain," *Neuropsychology* 19 (2005): 593–96.

17. Ibid., p. 597.

18. Michael Tomasello, *The Cultural Origins of Human Cognition* (Cambridge, MA: Harvard University Press, 1999), p. 22.

19. Ibid.

20. Ibid.

21. Gazzaniga, *The Mind's Past*, p. 153.

22. Ibid., p. 154.

23. Lev S. Vygotsky, *Thought and Language*. Translation newly revised and edited by Alex Kozulin (Cambridge, MA: MIT Press, 1986), p. 14.

24. David C. Rubin, "The Basic-Systems Model of Episodic Memory," *Perspectives on Psychological Science* 1 (2006): 284.

25. Eric Klinger, *Daydreaming: Using Waking Fantasy and Imagery for Self-knowledge and Creativity* (Los Angeles, CA: Jeremy P. Tarcher, 1990), pp. 68–69.

26. Ibid., p. 68.

27. Helen Markus and Elissa Wurf "The Dynamic Self-Concept: A Social and Psychological Perspective" *Annual Review of Psychology* 38 (1987): 299–301.

28. David J. Turk et al. "Mike or Me? Self-Recognition in a Split-Brain Patient" *Nature Neuroscience* 9 (September 2002): 841–42.

29. Mark L. Howe, "Memories from the Cradle," *Current Directions in Psychological Science* 12 (2003): 63.

30. Ibid., pp. 62–65.

31. Markus and Wurf, "The Dynamic Self-Concept," p. 304.

32. Shelley E. Taylor, *Positive Illusions: Creative Self-Deception and the Healthy Mind* (New York: Basic Books, 1989), p. 8.

33. Ibid., pp. 29–32.

34. Ibid., pp. 32–45.

35. Ibid., pp. 212–15.

36. Ibid., pp. 126–33.

37. Jill Bolte Taylor, *My Stroke of Insight: A Brain Scientist's Personal Journey* (New York: Viking, 2006), p. 66–67.

CHAPTER 6. MENTAL TIME TRAVEL

1. Endel Tulving, "Episodic Memory: From Mind to Brain," *Annual Review of Psychology* 53 (2002): 1–2.

2. Ronald T. Kellogg, *Cognitive Psychology*, 2nd ed. (Thousand Oaks: CA, 2003), p. 151.

3. Brenda Milner, "Amnesia following Operations on the Temporal Lobes." In *Amnesia*, eds. C. M. W. Whitty and O. L. Zangwill (London: Butterworths, 1966), pp. 112–14.

4. Ibid., pp. 113–14.

5. Tulving, "Episodic Memory," pp. 12–16.

6. Ibid., p. 14.

7. Endel Tulving, "Episodic Memory and Autonoesis: Uniquely Human?" In *The Missing Link in Cognition*, eds. Herbert S. Terrace and Janet Metcalfe (Oxford, Oxford University Press, 2005), p. 36.

8. Ibid., pp. 37–38.

9. Bennett L. Schwartz, "Do Nonhuman Primates Have Episodic Memory?" In *The Missing Link in Cognition*, eds. Herbert S. Terrace and Janet Metcalfe (Oxford, Oxford University Press, 2005), p. 233–35.

10. Tulving, "Episodic Memory and Autonoesis," p. 40.

11. Kellogg, *Cognitive Psychology*, p. 176.

12. Ulrich Neisser, "John Dean's Memory." In *Memory Observed: Remembering in Natural Contexts* 2nd ed. (New York: Worth Publishers, 2000), p. 272.

13. Ibid., p. 284.

14. M. A. Conway, "A Structural Model of Autobiographical Memory." In *Theoretical Perspectives on Autobiographical Memory*, eds. A. Conway, D. C. Rubin, H. Spinnler, and W. A. Wagenaar (Dordrecht, Netherlands: Kluwer, 1992), pp. 167–93.

15. D. Stephen Lindsay et al., "True Photographs and False Memories," *Psychological Science* 15 (2004): 150–53.

16. Michael S. Gazzaniga, *The Mind's Past* (Berkeley, University of California Press, 1998), p. 145.

17. Ibid., p. 146.

18. Karl K. Szpunar, Jason M. Watson, and Kathleen B. McDermott, "Neural Substrates of Envisioning the Future," *Proceedings of the National Academy of Science* 104 (January 2007): 645.

19. Ibid., p. 644.

20. Ibid.

21. Ibid., p. 645.

22. Randy L. Buckner, Jessica R. Andrews-Hanna, and Daniel L. Schacter, "The Brain's Default Network: Anatomy, Function, and Relevance to Disease," *Annals of the New York Academy of Sciences* 1124 (2008): 5.

23. Ibid., pp. 18–19.

24. Marcia K. Johnson, "Memory and Reality," *American Psychologist* 61 (November 2006): 760.

25. Ibid., p. 762.

26. Ibid., p. 765.

27. Nicholas P. Spanos, *Multiple Identities and False Memories: A Sociocognitive Perspective* (Washington, DC: American Psychological Association, 1996), pp. 77–79.

28. Ibid., pp. 98–103.

29. Elizabeth F. Loftus and Katherine Ketcham, *The Myth of Repressed Memory* (New York: St. Martin's Press, 1994), p. 26.

30. Daniel M. Bernstein and Elizabeth F. Loftus, "How to Tell If a Particular Memory Is True or False," *Perspectives on Psychological Science* 4 (2009): 371.

31. Kellogg, *Cognitive Psychology*, p. 200.

32. Loftus and Ketcham, *The Myth of Repressed Memory*, p. 58.

33. Spanos, *Multiple Identities and False Memories*, p. 117.

34. Giuliana A. L. Mazzoni, Elizabeth L. Loftus, and Irving Kirsch, "Changing Beliefs about Implausible Autobiographical Events: A Little Implausibility Goes a Long Way," *Journal of Experimental Psychology: Applied* 7 (2001): 57.

35. Spanos, *Multiple Identities and False Memories*, pp. 119–27.

36. T. D. Borkovec, William J. Ray, and Joachim Stöber, "Worry: A Cognitive Phenomenon Intimately Linked to Affective, Physiological, and Interpersonal Problems," *Cognitive Therapy and Research* 22 (1998): 562–63.

37. Constantine Sedikides, Tim Wildschut, Jamie Arndt, and Clay Routledge, "Nostalgia: Past, Present, and Future," *Current Directions in Psychological Science* 17 (2008): 305.

38. Ibid.

39. Susan Nolen-Hoeksema, Blair E. Wisco, and Sonja Lyubomirsky, "Rethinking Rumination," *Perspectives on Psychological Science* 3 (2008): 400.

CHAPTER 7. EMOTIONS

1. Antonio R. Damasio, *The Feeling of What Happens: Body and Emotion in the Making of Consciousness* (New York: Harcourt Brace & Company, 1999), pp. 59–62.

2. Ibid., pp. 53–55.

3. Carl Izard, "Emotion Theory and Research: Highlights, Unanswered Questions, and Emerging Issues," *Annual Review of Psychology* 60 (2009): 15.

4. Richard S. Lazarus, "Thoughts on the Relation between Emotion and Cognition," *American Psychologist* 37 (1982): 1023.

5. Ibid., p. 1024.

6. Joseph E. LeDoux, "Emotion, Memory, and the Brain," *Scientific American* 12, no. 1 (2002): pp. 64–70.

7. Ibid., p. 69.

8. Ibid.

9. Ibid., p. 70.

10. Joseph E. LeDoux, "Emotion Circuits in the Brain," *Annual Review of Neuroscience* 23 (2000): 175.

11. T. D. Borkovec, William J. Ray, and Joachim Stöber, "Worry: A Cognitive Phenomenon Intimately Linked to Affective, Physiological, and Interpersonal Problems," *Cognitive Therapy and Research* 22 (1998): 562–63.

12. Anthony Charuvastra and Marylene Cloitre," Social Bonds and Posttraumatic Stress Disorder," *Annual Review of Psychology* 59 (2008): 304.

13. Thomas Childers, *Soldier from the War Returning: The Greatest Generation's Troubled Homecoming from World War II* (Boston: Houghton Mifflin Harcourt, 2009), p. 17.

14. Ibid., pp. 254–74.

15. Ibid., p. 135.

16. John L. Cotton, "A Review of Research on Schacter's Theory of Emotion and the Misattribution of Arousal," *European Journal of Social Psychology* 11 (1981): 366.

17. Ibid., p. 374.

18. Shelley E. Taylor and Jonathan D. Brown, "Illusion and Well-Being: A Social Psychological Perspective on Mental Health," *Psychological Bulletin* 103 (1988): 193.

19. Jutta Joorman, "Cognitive Inhibition and Emotion Regulation in Depression," *Current Directions in Psychological Science* 19 (2010): 161–63.

20. Thomas F. Oltmanns and Robert E. Emery, *Abnormal Psychology*, 4th ed. (Upper Saddle River, NJ: Pearson Education, 2004), pp. 83–84.

21. Elizabeth A. Phelps, "Emotion and Cognition: Insights from the Study of the Human Amygdala," *Annual Review of Psychology* 2006: 44–45.

22. Ibid., p. 44.

23. Sara W. Lazar et al., "Functional Brain Mapping of the Relaxation Response and Meditation," *NeuroReport* 11, no. 7 (2000): 1582.

24. Ibid.

25. Herbert Benson and William Proctor, *Relaxation Revolution: Enhancing Your Personal Health through the Science and Genetics of Mind-Body Healing* (New York: Scribner, 2010), p. 9.

26. Antoine Lutz, Heleen A. Slagter, John D. Dunne, and Richard J. Davidson, "Attention Regulation and Monitoring in Meditation," *Trends in Cognitive Science* 12, no. 4 (2008): 163.

27. Shelley E. Taylor, *Positive Illusions: Creative Self-Deception and the Healthy Mind* (New York: Basic Books, 1989), pp. 115–20.

28. Ibid., p. 119.

29. Ibid.

30. Donald D. Price, Damien G. Finniss, and Fabrizio Benedetti, "A Comprehensive Review of the Placebo Effect: Recent Advances and Current Thought," *Annual Review of Psychology* 59 (2008): 568–69.

31. Steven E. Hyman, Robert C. Malenka, and Eric J. Nestler, "Neural Mechanisms of Addiction: The Role of Reward-Related Learning and Memory," *Annual Review of Neuroscience* 32 (2006): 571–72.

32. George F. Koop and Michel Le Moal, "Addiction and the Brain Antireward System," *Annual Review of Psychology* 59 (2008): 32.

33. Terry E. Robinson and Kent C. Berridge, "Addiction," *Annual Review of Psychology* 54 (2003): 34.

34. Eva Kemps and Marika Tiggemann, "A Cognitive Experimental Approach to Understanding and Reducing Food Cravings," *Current Directions in Psychological Science* 19 (2010): 86.

35. Ibid., p. 87.

36. Christine Harris, "The Evolution of Jealousy," *American Scientist* 92 (2004): 62.

37. Ibid., p. 66.

38. Christine R. Harris, "A Review of Sex Differences in Sexual Jealousy, Including Self-Report Data, Psychophysiological Responses, Interpersonal Violence, and Morbid Jealousy," *Personality and Social Psychology Review* 7 (2003): 115.

39. Harris, "Evolution of Jealousy," p. 70.

CHAPTER 8. THE SOCIAL MIND

1. John Bartlett, *Familiar Quotations: A Collection of Passages, Phrases, and Proverbs Traced to Their Source in Ancient and Modern Literature*, 16th ed., ed. Justin Kaplan (Boston: Little, Brown, and Company, 1992), p. 231.

2. Gregory Carey, *Human Genetics for the Social Sciences* (Thousand Oaks, CA: Sage Publications, 2003), p. 237.

3. Roy F. Baumeister and Mark R. Leary, "The Need to Belong: Desire for Interpersonal Attachments as a Fundamental Human Motivation," *Psychological Bulletin* 117 (1995): 497.

4. Ibid., p. 503.

5. Naomi I. Eisenberger, Matthew D. Lieberman, and Kipling D. Williams, "Does Social Rejection Hurt? An fMRI Study of Social Exclusion," *Science* 302 (2003): 291.

6. Ibid.

7. Shelley E. Taylor, "Tend and Befriend: Biobehavioral Bases of Affiliation under Stress," *Current Directions in Psychological Science* 15 (2006): 273.

8. Ibid., p. 274.

9. Baumeister and Leary, "The Need to Belong," p. 499.

10. Leslie C. Aiello and R. I. M. Dunbar, "Neocortex size, Group Size, and the Evolution of Language," *Current Anthropology* 34 (April 1993): 185.

11. Ibid.

12. Ibid., p. 184.

13. Susan T. Fiske, "Social Cognition and the Normality of Prejudgment." In *On the Nature of Prejudice: Fifty Years after Allport*, eds. John F. Dovidio, Peter Glick, and Laurie A. Rudman (Oxford: Blackwell Publishing, 2005), pp. 37–40.

14. Matthew D. Lieberman "Social Cognitive Neuroscience: A Review of Core Processes," *Annual Review of Psychology* 58 (2007): 272.

15. Mary E. Wheeler and Susan T. Fiske, "Controlling Racial Prejudice: Social-Cognitive Goals Affect Amygdala and Stereotype Activation," *Psychological Science* 16 (2005): 56–60.

16. Rebecca S. Bigler and Lynn S. Liben, "Developmental Intergroup Theory: Explaining and Reducing Children's Social Stereotyping and Prejudice," *Current Directions in Psychological Science* 16 (2007): 162–63.

17. Mark Mazower, *Hitler's Empire: How the Nazis Ruled Europe* (New York: Penguin Press, 2008), pp. 182–83.

18. Anthony G. Greenwald, T. Andrew Poehlman, Eric Luis Uhlmann, and Jahzarin R. Banaji, "Understanding and Using the Implicit Association Test: III. Meta-Analysis of Predictive Value," *Journal of Personality and Social Psychology* 97 (2009): 17–41.

19. Mazower, *Hitler's Empire*, p. 586.

20. Saul Kassin, Steven Fein, Hazel Rose Markus, *Social Psychology*, 7th ed. (Boston: Houghton Mifflin Company 2008), pp. 109–10.

21. Douglas S. Krull et al. "The Fundamental Attribution Error: Correspondence Bias in Individualist and Collectivist Cultures," *Personality and Social Psychology Bulletin* 25 (1999): 1208.

22. Ibid., pp. 1211–12.

23. Stephen Jay Gould, *The Mismeasure of Man*, revised and expanded (New York: W. W. Norton & Company, 1996), p. 104.

24. Ibid., p. 106.

25. Ibid., p. 78.

26. Alan R. Templeton, "Human Races: A Genetic and Evolutionary Perspective," *American Anthropologist* 100 (1999): 633.

27. Ibid.

28. Gould, *The Mismeasure of Man*, p. 97.

29. Ibid., p. 92.

30. Ibid., pp. 135–36.

31. Amy J. C. Cuddy et al., "Stereotype Content Model across Cultures: Towards Universal Similarities and Some Differences," *British Journal of Social Psychology* 48 (2009): 2–5.

32. Ibid., p. 6.

33. Ibid., p. 4.

34. Mina Cikara, Rachel A. Farnsworth, Lasana T. Harris, and Susan T. Fiske, "On the Wrong Side of the Trolley Track: Neural Correlates of Relative Social Evaluation," *Scan* 5 (2010): 405.

35. Ibid., p. 409.

36. Ibid., p. 410–11.

CHAPTER 9. MORALITY

1. Marvin W. Berkowitz and Stephen A. Sheldon, "Fairness." In *Character Strengths and Virtues: A Handbook and Classification*, eds. Christopher Peterson and Martin E. P. Seligman (New York: American Psychological Association and Oxford University Press, 2004), pp. 391–92.

2. Ibid., pp. 392–93.

3. Ibid., pp. 394–95.

4. Ibid., pp. 396–97.

5. Ibid., p. 399.

6. Bruno S. Frey, David A. Savage, and Benno Torgler, "Interaction of Natural Survival Instincts and Internalized Social Norms Exploring the *Titanic* and *Lusitania* Disasters," *Proceedings of the National Academy of Sciences* 107 (2010): 4862–65.

7. Ibid., p. 4863.

8. Albert Bandura, "Moral Disengagement in the Perpetration of Inhumanities," *Personality and Social Psychology Review* 3 (1999): 196.

9. Jerry M. Burgher, "Replicating Milgram: Would People Still Obey Today?" *American Psychologist* 64 (2009): 1.

10. Ibid., p. 8.

11. Ibid., p. 1.

12. Joshua D. Greene, R. Brian Sommerville, Leigh E. Nystrom, John M. Darley, and Jonathon D. Cohen, "An fMRI Investigation of Emotional Engagement in Moral Judgment," *Science* 293 (September 2001): 2106.

13. Ibid.

14. Randy L. Buckner, Jessica R. Andrews-Hanna, and Daniel L. Schacter, "The Brain's Default Network: Anatomy, Function, and Relevance to Disease," *Annals of the New York Academy of Sciences* 1124 (2008): 21.

15. Jonathan Haidt, "Morality," *Perspectives on Psychological Science* 3 (2008): 69.

16. Jana Schaich Borg, Debra Lieberman, and Kent A. Kiehl, "Infection, Incest, and Iniquity: Investigating the Neural Correlates of Disgust and Morality," *Journal of Cognitive Neuroscience* 20, no. 9 (2008): 1530.

17. Ibid., p. 1541.

18. Marco Iacoboni et al., "Grasping the Intentions of Others with One's Own Mirror Neuron System," *PLoS Biology* 3 e79 (March 2005): 0531.

19. Chadd M. Funk and Michael S. Gazzaniga, "The Functional Brain Architecture of Human Morality," *Current Opinion in Neurobiology* 19 (2009): 679.

20. Ibid., p. 680.

21. Thalia Wheatley and Jonathan Haidt, "Hypnotic Disgust Makes Moral Judgments More Severe," *Psychological Science* 16 (2005): 781.

22. Ibid., pp. 782–83.

23. Ibid., p. 783.

24. Funk and Gazzaniga, "The Functional Brain Architecture of Human Morality," p. 680.

25. Iacoboni et al., "Grasping the Intentions of Others," p. 0529.

26. Jean Decety and Andrew N. Meltzoff, "Empathy, Imitation, and the Social Brain." In *Empathy: Philosophical and Psychological Perspectives*, eds. Amy Copland and Peter Goldie (New York: Oxford University Press, 2011), pp. 68–69.

27. Ibid., pp. 69–70.

28. Jean Decety and Phillip L. Jackson, "A Social-Neuroscience Perspective on Empathy," *Current Directions in Psychological Science* 15 (2006): 54.

29. Ibid., p. 55.

30. Ibid.

31. Iacoboni et al. "Grasping the Intentions of Others," p. 0531.

32. Decety and Jackson, "A Social-Neuroscience Perspective on Empathy," p. 56.

33. Ibid., pp. 56–57.

34. Roy Baumeister and Julie Juola Exline, "Virtue, Personality, and Social Relations: Self-Control as the Moral Muscle," *Journal of Personality* 67 (1999): 1176.

35. Cynthia L. Ogden et al., "Prevalence of Obesity in the United States, 2009–2010," NCHS Data Brief No. 82, available from http://www.cdc.gov/nchs/data/databriefs/db82.pdf (accessed January 2012), p. 1.

36. Nouriel Roubini and Stephen Mihm, *Crisis Economics: A Crash Course in the Future of Finance* (New York: Penguin Press, 2010), p. 32.

37. Natasha Vargas-Cooper, "Hard Core: The New World of Porn Is Revealing Eternal Truths about Men and Women" *Atlantic* (January/February 2011), p. 97.

38. Thomas Suddendorf and Michael C. Corballis, "The Evolution of Foresight: What Is Mental Time Travel, and Is It Unique to Humans?" *Behavioral and Brain Sciences* 30 (June 2007): 312.

39. Baumeister and Exline, "Virtue, Personality, and Social Relations," p. 1177.

40. Ibid.

41. Ibid.

42. Ibid., p. 1178.

43. Annabelle Belcher and Walter Sinnott-Armstrong, "Neurolaw," *WIREs Cognitive Science* 1 (January/February 2010): 19.

44. Ibid., pp. 19–20.

45. Laurence Steinberg, "Risk Taking in Adolescence: New Perspectives from Brain and Behavioral Science," *Current Directions in Psychological Science* 16 (2007): 56.

46. Ibid.

47. Ibid.

48. Michael Koenigs et al., "Damage to the Prefrontal Cortex Increases Utilitarian Moral Judgments," *Nature* 446 (2007): 908.

49. Michael F. Lorber, "Psychophysiology of Aggression, Psychopathy, and Conduct Problems: A Meta-Analysis," *Psychological Bulletin* 130 (2004): 532.

CHAPTER 10. SPIRITUALITY

1. Andrew Curry, "Gobekli Tepe: The World's First Temple?" *Smithsonian Magazine*, http://www.smithsonianmag.com/history-archaeology/gobekli-tepe.html?c=y&page=3 (accessed November 15, 2011).

2. Robert A. Emmons and Raymond F. Paloutzian, "The Psychology of Religion," *Annual Review of Psychology* 54 (2003): 379.

3. Krista Tippett, *Speaking of Faith* (New York: Viking Press, 2007), pp. 231–32.

4. William James, *Varieties of Religious Experience: A Study in Human Nature* (New York: Collier Books, 1961), p. 59.

5. Ibid., pp. 61–62.

6. Sheldon Solomon, Jeff Greenberg, and Tom Pyszczynski, "Pride and Prejudice: Fear of Death and Social Behavior," *Current Directions in Psychological Science* 9 (2000): 200.

7. Ibid., p. 201.

8. Sander L. Koole, Jeff Greenberg, and Tom Pyszczynski, "Introducing Science to the Psychology of the Soul: Experimental Existential Psychology," *Current Directions in Psychological Science* 15 (2006): 213–14.

9. Ibid., p. 214.

10. Ibid.

11. Emmons and Paloutzian, "The Psychology of Religion," p. 383.

12. Kevin S. Seybold and Peter C. Hill, "The Role of Religion and Spirituality in Mental and Physical Health," *Current Directions in Psychological Science* 10 (2001): 22.

13. Ibid.

14. Ibid., pp. 22–23.

15. Eva Jonas and Peter Fischer, "Terror Management and Religion: Evidence That Intrinsic Religiousness Mitigates Worldview Defense Following Mortality Salience," *Journal of Personality and Social Psychology* 91 (2006): 557.

16. Ibid., pp. 558–59.

17. Ibid., p. 563.

18. Emmons and Paloutzian, "The Psychology of Religion," pp. 387–88.

19. Ibid., pp. 388–89.

20. Michael Inzlicht et al., "Neural Markers of Religious Conviction," *Psychological Science* 20 (2009): 385–87.

21. Ibid., p. 388.

22. Ibid., p. 389.

23. Michael E. McCullough and Brian L. B. Wiloughby, "Religion, Self-Regulation, and Self-Control: Associations, Explanations, and Implications," *Psychological Bulletin* 135 (2009): 70–72.

24. Ibid., p. 82.

25. Ibid., p. 87.

26. James, *The Varieties of Religious Experience*, pp. 299–301.

27. Ibid., p. 300.

28. Ibid., p. 301.

29. Ibid., p. 323.

30. Andrew Newberg, Eugene G. d'Aquili and Vince Rause, *Why God Won't Go Away: Brain Science and Biology of Believing* (New York: Ballantine Books, 2001), p. 110.

31. James, *The Varieties of Religious Experience*, p. 29.

32. Ibid., p. 30.

33. Newberg, D'Aquili, and Rause, *Why God Won't Go Away*, p. 112.

34. Ibid.

35. Ibid., p. 110.

36. Ibid., p. 80.

37. Ibid., pp. 28–29.

38. Ibid., pp. 3–7.

39. Niels G. Waller, Brian A. Kojetin, Thomas J. Bouchard Jr., David T. Lykken, and Auke Tellegen, "Genetic and Environmental Influences on Religious Interests, Attitudes, and Values: A Study of Twins Reared Apart and Reared Together," *Psychological Science* 1 (1990): 138–41.

40. Ibid., p. 138.

41. Jesse M. Bering, "The Cognitive Psychology of Belief in the Supernatural," *American Scientist* 94 (2006): 142–49.

42. Ibid., p. 144.

43. Dimitrios Kapogiannis, Aron K. Barbey, Michael Su, Giovanna Zamboni, Frank Krueger, and Jordan Grafman, "Cognitive and Neural Foundations of Religious Belief," *Proceedings of the National Academy of Sciences* 106 (March 2009): 4876–81.

44. Ibid., p. 4876.

45. Ibid., p. 4879.

46. John Micklethwait and Adrian Wooldridge, *God Is Back: How the Global Revival of Faith Is Changing the World* (New York: Penguin Press, 2009), pp. 9–10.

47. Ibid., p. 16.

48. Ibid., pp. 4–5.

49. Stephen Jay Gould, *Rock of Ages: Science and Religion in the Fullness of Life* (New York: Ballantine Publishing Group, 1999), p. 188.

50. Francis S. Collins, *The Language of God: A Scientist Presents Evidence for Belief* (New York: Free Press, 2006), p. 107.

51. Gregory W. Graffin and William B. Provine, "Evolution, Religion, and Free Will," *American Scientist* 95 (July–August 2007): 296.

52. Gould, *Rock of Ages*, pp. 5–6.

53. Ibid., p. 54.

54. Ibid., p. 55.

55. Antony Flew, *There is a God: How the World's Most Notorious Atheist Changed His Mind* (New York: HarperCollins, 2007), pp. 136–38.

56. Ibid., p. 138.

57. Karen Armstrong, *A History of God: The Four Thousand Year Quest of Judaism, Islam, and Christianity* (New York: Random House 1993), p. xix.

CHAPTER 11. TWENTY-FIRST-CENTURY MIND

1. Luigi Luca Cavalli-Sforza and Francesco Cavalli-Sforza, *The Great Human Diasporas: The History of Diversity and Evolution* (Reading, MA: Addison Wesley Publishing Company, 1995), p. 121.

2. Ibid., p. 123.

3. Margaret W. Conkey, "A History of the Interpretation of European 'Paleolithic Art': Magic, Mythogram, and Metaphors for Modernity." In *Handbook of Human Symbolic Evolution*, eds. Andrew Lock and Charles R. Peters (Oxford: Clarendon Press, 1996), pp. 288–95.

4. Michael Fitzgerald, "Portals to Other Realities: Legend Rock Carries 10,000 Years of Profound Beliefs," *Wall Street Journal*, Leisure & Arts (September 18–19, 2010), p. W14.

5. Lev Vygotsky, *Mind in Society: The Development of Higher Psychological Processes*, eds. Michael Cole, Vera John-Steiner, Sylvia Scribner, Ellen Souberman (Cambridge, MA: Harvard University Press, 1978), p. 125.

6. Deborah Schmandt-Bessarat, "Tokens as Precursors of Writing." In *Writing: A Mosaic of New Perspectives*, eds. Elena L. Grigorenko, Elisa Mambrino, and David D. Preiss (New York: Psychology Press, 2012), p. 3.

7. David R. Olson, *The World on Paper: The Conceptual and Cognitive Implications of Writing and Reading* (Cambridge: Cambridge University Press, 1994), pp. 84–85.

8. Michael I. Posner and Mary K. Rothbart, *Educating the Human Brain* (Washington, DC: American Psychological Association, 2007), pp. 150–51.

9. Ibid., pp. 152–53.

10. Ibid., p. 154.

11. Ibid., pp. 176–79.

12. Ibid., pp. 181–82.

13. Merlin Donald, *Origins of the Modern Mind: Three Stages in the Evolution of Culture and Cognition* (Cambridge, MA: Harvard University Press, 1991), p. 287.

14. Jerome Bruner, *Actual Minds, Possible Worlds* (Cambridge, MA: Harvard University Press, 1986), p. 11.

15. Donald, *Origins of the Modern Mind*, p. 273.

16. Ibid.

17. Ibid., pp. 273–74.

18. Ibid., p. 335.

19. Ronald T. Kellogg, "Training Writing Skills: A Cognitive Developmental Perspective," *Journal of Writing Research* 1 (2008): 4–7.

20. Walter Ong, *Orality and Literacy: The Technologizing of the Word* (London: Methuen, 1982), p. 172.

21. Olson, *The World on Paper*, p. 43.

22. Ong, *Orality and Literacy*, p. 79.

23. David Rubin, *Memory in Oral Traditions: The Cognitive Psychology of Epic, Ballads, and Counting-out Rhymes* (New York: Oxford University Press, 1995), p. 139.

24. Michelle Eskrit, Kang Lee, and Merlin Donald, "The Influence of Symbolic Literacy on Memory: Testing Plato's Hypothesis," *Canadian Journal of Experimental Psychology* 55 (2001): 41–43.

25. W. Michael Cox and Richard Alm, "You Are What You Spend," *New York Times, Sunday Opinion*, February 10, 2008, final edition, p. 14.

26. Ibid.

27. Emory H. Woodard IV, *Media in the Home 2000: The Fifth Annual Survey of Parents and Children* (Philadelphia, PA: Annenberg Public Policy Center of the University of Pennsylvania, Survey Series no. 7), p. 8.

28. Ibid., p. 19.

29. Dorothy G. Singer and Jerome R. Singer, *Imagination and Play in the Electronic Age* (Cambridge, MA: Harvard University Press, 2005), pp. 63–64.

30. Patti M. Valkenburg and Joehen Peter, "Social Consequences of the Internet for Adolescents: A Decade of Research," *Current Directions in Psychological Science* 18 (2009): 1–2.

31. Nicholas Carr, *The Shallows* (New York: W. W. Norton & Company, 2010), p. 6.

32. Ibid., p. 10.

33. Maggie Jackson, *Distracted: The Erosion of Attention and the Coming Dark Age* (Amherst, NY: Prometheus Books, 2008), p. 73.

34. Hee Jeong Yoo et al., "Attention Deficit Hyperactivity Symptoms and Internet Addiction," *Psychiatry and Clinical Neurosciences* 58 (2004): 487–94.

35. Carl Erik Landhuis, Richie Poulton, David Welch, and Robert John Hancox, "Does Television Viewing Lead to Attention Problems in Adolescence? Results from a Prospective Longitudinal Study," *Pediatrics* 120 (2007): 533–36.

36. Edward L. Swing, David A. Gentile, Craig A. Anderson, and David A. Walsh, "Television and Video Game Exposure and the Development of Attention Problems," *Pediatrics* 126 (2010): 216–18.

37. Ibid., p. 220.

38. Stephen Marche, "The Book That Contains All Books," *Wall Street Journal* (October 17–18, 2009), p. W9.

39. Matt Ridley, "Humans: Why They Triumphed," *Wall Street Journal* (May 22–23, 2010), pp. W1–W2.

40. Don Tapscott and Anthony D. Williams, *Wikinomics: How Mass Collaboration Changes Everything* (New York: Portfolio, 2006), p. 9.

41. Michael Nielsen, *Reinventing Discovery* (Princeton, NJ: Princeton University Press, 2012), p. 10.

42. Seth Cooper et al., "Predicting Protein Structures with a Multiplayer Online Game," *Nature* 466 (2010): 756–60.

BIBLIOGRAPHY

Aiello, Leslie C., and R. I. M. Dunbar. "Neocortex Size, Group Size, and the Evolution of Language." *Current Anthropology* 34 (April 1993): 184–93.

Armstrong, Karen. *A History of God: The Four Thousand Year Quest of Judaism, Islam, and Christianity*. New York: Random House, 1993.

Aitchison, Jean. *The Seeds of Speech: Language Origin and Evolution*. Cambridge: Cambridge University Press, 1996.

Baddeley, Alan D. *Working Memory*. New York: Oxford University Press, 1986.

Baddeley, Alan, Susan Gathercole, and Costanza Papagno. "The Phonological Loop as a Language Learning Device." *Psychological Review* 105 (1998): 158–73.

Bandura, Albert. "Moral Disengagement in the Perpetration of Inhumanities." *Personality and Social Psychology Review* 3 (1999): 193–209.

Baron-Cohen, Simon, Alan. M. Leslie, Uta Frith. "Does the Autistic Child Have a 'Theory of Mind.'" *Cognition* 21 (1985): 37–46.

Bartlett, John. *Familiar Quotations: A Collection of Passages, Phrases, and Proverbs Traced to Their Source in Ancient and Modern Literature*, 16th ed. Edited by Justin Kaplan. Boston: Little, Brown, and Company, 1992.

Baumeister, Roy F., and Julie Juola Exline. "Virtue, Personality, and Social Relations: Self-Control as the Moral Muscle." *Journal of Personality* 67 (1999): 1165–94.

Baumeister, Roy F. and Mark R. Leary. "The Need to Belong: Desire for Interpersonal Attachments as a Fundamental Human Motivation." *Psychological Bulletin* 117 (1995): 497–529.

Beatty, Jackson. *The Human Brain: Essentials of Neuroscience*. Thousand Oaks, CA: Sage Publications, 2001.

Belcher, Annabelle, and Walter Sinnott-Armstrong. "Neurolaw." *WIREs Cognitive Science* 1 (January/February 2010): 18–22.

Benson, Herbert, and William Proctor. *Relaxation Revolution: Enhancing Your Personal Health through the Science and Genetics of Mind-Body Healing*. New York: Scribner, 2010.

Berkowitz, Marvin W., and Stephen A. Sheldon. "Fairness." In *Character Strengths and Virtues: A Handbook and Classification*. Edited by Christopher Peterson and Martin E. P. Seligman. New York: American Psychological Association and Oxford University Press, 2004, pp. 391–412.

Bernstein, Daniel M., and Elizabeth F. Loftus. "How to Tell If a Particular Memory Is True or False," *Perspectives on Psychological Science* 4 (2009): 371–74.

Bering, Jesse M. "The Cognitive Psychology of Belief in the Supernatural." *American Scientist* 94 (2006): 142–49.

Bickerton, Derek. *Language and Species*. Chicago: University of Chicago Press, 1990.

Bigler, Rebecca S., and Lynn S. Liben. "Developmental Intergroup Theory: Explaining and Reducing Children's Social Stereotyping and Prejudice." *Current Directions in Psychological Science* 16 (2007): 162–66.

Borkovec, T. D., William J. Ray, and Joachim Stöber. "Worry: A Cognitive Phenomenon Intimately Linked to Affective, Physiological, and Interpersonal problems." *Cognitive Therapy and Research* 22 (1998): 561–76.

Borg, Jana Schaich, Debra Lieberman, and Kent A. Kiehl. "Infection, Incest, and Iniquity: Investigating the Neural Correlates of Disgust and Morality." *Journal of Cognitive Neuroscience* 20, no. 9 (2008): 1529–46.

Brooks, Rechele, and Andrew N. Meltzoff, "The Development of Gaze Following and Its Relation to Language." *Developmental Science* 8 (2005): 535–43.

Bruner, Jerome. *Actual Minds, Possible Worlds*. Cambridge, MA: Harvard University Press, 1986.

Buckner, Randy L., Jessica R. Andrews-Hanna, and Daniel L. Schacter. "The Brain's Default Network: Anatomy, Function, and Relevance to Disease." *Annals of the New York Academy of Sciences* 1124 (2008): 1–38.

Burgher, Jerry M. "Replicating Milgram: Would People Still Obey Today?" *American Psychologist* 64 (2009): 1–11.

Carey, Gregory. *Human Genetics for the Social Sciences*. Thousand Oaks, CA: Sage Publications, 2003.

Carr, Nicholas. *The Shallows*. New York: W. W. Norton & Company, 2010.

Cavalli-Sforza, Luigi Luca, Alberto Piazza, Paolo Menozzi, and Joanna Mountain. "Reconstruction of Human Evolution: Bringing Together Genetic, Archaeological, and Linguistic Data. *Proceedings of the National Academy of Sciences* 85 (1988): 6002–6006.

Cavalli-Sforza, Luigi Luca and Francesco Cavalli-Sforza. *The Great Human Diasporas: The History of Diversity and Evolution*. Reading, MA: Addison Wesley Publishing Company, 1995.

Carroll, Sean B. "Genetics and the Making of *Homo sapiens*." *Nature* 422 (April 2003): 849–57.

Charuvastra, Anthony, and Marylene Cloitre. "Social Bonds and Posttraumatic Stress Disorder. *Annual Review of Psychology* 59 (2008): 301–28.

Childers, Thomas. *Soldier from the War Returning: The Greatest Generation's Troubled Homecoming from World War II*. Boston: Houghton Mifflin Harcourt, 2009.

Christiansen, Morton H. and Simon Kirby. "Language Evolution: Consensus and Controversies." *TRENDS in Cognitive Sciences* 7 (July 2003): 301–307.

Cikara, Mina, Rachel A. Farnsworth, Lasana T. Harris, and Susan T. Fiske. "On the Wrong Side of the Trolley Track: Neural Correlates of Relative Social Evaluation." *Scan* 5 (2010): 404–13.

Clark, Herbert H. *Using Language*. Cambridge: Cambridge University Press, 1996.

Cochran, Gregory, and Henry Harpending. *The 10,000 Year Explosion: How Civilization Accelerated Human Evolution*. New York: Basic Books, 2009.

Collins, Francis S. *The Language of God: A Scientist Presents Evidence for Belief*. New York: Simon and Schuster, 2006.

Conkey, Margaret W. "A History of the Interpretation of European 'Paleolithic Art': Magic, Mythogram, and Metaphors for Modernity." In *Handbook of Human Symbolic Evolution*. Edited by Andrew Lock and Charles R. Peters. Oxford: Clarendon Press, 1996, pp. 288–350.

Conway, M. A. "A Structural Model of Autobiographical Memory," In *Theoretical Perspectives on Autobiographical Memory*." Edited by M. A. Conway, D. C. Rubin, H. Spinnler, and W. A. Wagenaar. Dordrecht, Netherlands: Kluwer, 1992.

Cooper, Seth, Firas Khatib, Adrien Reuille, Janos Barbero, Jeehyung Lee, Michael Beenen, Andrew Leaver-Fay, David Baker, Zoran Popovi and Foldit Players. "Predicting Protein Structures with a Multiplayer Online Game." *Nature* 466 (2010): 756–60.

Cotton, John L. "A Review of Research on Schacter's Theory of Emotion and the Misattribution of Arousal." *European Journal of Social Psychology* 11 (1981): 365–97.

Cowan, Nelson. "The Magical Number 4 in Short-Term Memory: A Reconsideration of Mental Storage." *Behavioral and Brain Sciences* 24 (2001): 87–185.

Cox, W. Michael, and Richard Alm, "You Are What You Spend." *New York Times*, Sunday Opinion, February 10, 2008, final edition.

Cuddy, Amy J. C., Susan T. Fiske, Virginia S. Y. Kwan, Peter Glick, Stephanie Demoulin, Jacques-Phillip Leyens, Michael Harris Bond, et al. "Stereotype Content Model across Cultures: Towards Universal Similarities and Some Differences." *British Journal of Social Psychology* 48 (2009): 1–33.

Curry, Andrew. "Gobekli Tepe: The World's First Temple?" *Smithsonian Magazine*, November 2008, http://www.smithsonianmag.com/history-archaeology/gobekli-tepe.html?c=y&page =3 (accessed November 15, 2011).

Damasio, Antonio R. *The Feeling of What Happens: Body and Emotion in the Making of Consciousness*. New York: Harcourt Brace & Company, 1999.

Darwin, Charles. *The Descent of Man and Selection in Relation to Sex*. 2nd ed. New York: Collier, 1905.

Davis, Joel. *Mother Tongue: How Humans Create Language*. New York: Birch Lane Press, 1994.

Deacon, Terrence W. *The Symbolic Species: The Co-Evolution of Language and the Brain*. New York: W. W. Norton & Company, 1997.

Decety, Jean, and Andrew N. Meltzoff. "Empathy, Imitation, and the Social Brain." In *Empathy: Philosophical and Psychological Perspectives*. Edited by Amy Copland and Peter Goldie. New York: Oxford University Press, 2011, pp. 58–81.

Decety, Jean, and Phillip L. Jackson. "A Social-Neuroscience Perspective on Empathy." *Current Directions in Psychological Science* 15 (2006): 54–58.

Dennis, Carey. "Primate Evolution: Branching Out." *Nature* 437 (September 2005): 17–19.

Donald, Merlin. *Origins of the Modern Mind: Three Stages in the Evolution of Culture and Cognition.* Cambridge, MA: Harvard University Press, 1991.

Duncan, John, Rüdiger J. Seitz, Jonathan Kolodny, Daniel Bor, Hans Herzog, Ayesha Ahmed, Fiona N. Newell, and Hazel Emslie. "A Neural Basis for General Intelligence." *Science* 289 (July 2000): 457–60.

Eisenberg Naomi I., Matthew D. Lieberman, and Kipling D. Williams. "Does Social Rejection Hurt? An fMRI Study of Social Exclusion." *Science* 302 (2003): 291–92.

Emmons, Robert A., and Raymond F. Paloutzian. "The Psychology of Religion." *Annual Review of Psychology* 54 (2003): 377–402.

Enard, Wolfgang, Molly Przeworski, Simon E. Fisher, Celeilia S. L. Lai, Victor Wiebe, Takashi Kitano, Anthony P. Monaco, and Svante Pääbo. "Molecular Evolution of FOXP2, a Gene Involved in Speech and Language." *Nature* 418 (August 2002): 869–72.

Engle, Randall W., Stephen Tuholski, James E. Laughlin, and Andrew R. A. Conway. "Working Memory, Short-Term Memory, and General Fluid Intelligence: A Latent-Variable Approach." *Journal of Experimental Psychology: General* 128 (1999): 309–31.

Eskrit, Michelle, Kang Lee, and Merlin Donald. "The Influence of Symbolic Literacy on Memory: Testing Plato's Hypothesis." *Canadian Journal of Experimental Psychology* 55 (2001): 39–50.

Fagot, Joël and Carlo De Lillo. "A Comparative Study of Working Memory: Immediate Serial Recall in Baboons (*Papio papio*) and Humans." *Neuropsychologia* 49 (2011): 3870–80.

Fiske, Susan T. "Social Cognition and the Normality of Prejudgment." In *On the Nature of Prejudice: Fifty Years after Allport*. Edited by John F. Dovidio, Peter Glick, and Laurie A. Rudman. Oxford: Blackwell Publishing, 2005, pp. 36–53.

Fitzgerald, Michael "Portals to Other Realities: Legend Rock Carries 10,000 Years of Profound Beliefs." *Wall Street Journal*, Leisure & Arts (September 18–19, 2010), p. W14.

Flew, Antony. *There is a God: How the World's Most Notorious Atheist Changed His Mind*. New York: HarperCollins, 2007.

Fouts, Roger. *Next of Kin: What Chimpanzees Have Taught Me about Who We Are*. New York: William Morrow and Company, 1997.

Frey, Bruno S., David A. Savage, and Benno Torgler. "Interaction of Natural Survival Instincts and Internalized Social Norms Exploring the *Titanic* and *Lusitania* Disasters." *Proceedings of the National Academy of Sciences* 107 (2010): 4862–65.

Funk, Chadd M., and Michael S. Gazzaniga. "The Functional Brain Architecture of Human Morality" *Current Opinion in Neurobiology* 19 (2009): 678–81.

Gardner, Beatrice T., and R. Allen Gardner. "Evidence for Sentence Constituents in the Early Utterances of Child and Chimpanzee." *Journal of Experimental Psychology: General* 104 (1975): 244–67.

Gathercole, Susan E., and Alan D. Baddeley. *Working Memory and Language*. Hillsdale, NJ: Lawrence Erlbaum, 1993.

Gazzaniga, Michael S. "Cerebral Specialization and Interhemispheric Communication: Does the Corpus Callosum Enable the Human Condition?" *Brain* 123 (2000): 1293–1326.

———. "Forty-five Years of Split Brain Research and Still Going Strong" *Nature Reviews: Neuroscience* 6 (August 2005): 653–59.

———. *The Mind's Past.* Berkeley: University of California Press, 1998.

Gazzaniga, Michael S., Richard B. Ivry, and George R. Mangun, *Cognitive Neuroscience: The Biology of the Mind.* New York: W. W. Norton & Company, 1998.

Geary. David C. *The Origin of Mind: Evolution of Brain, Cognition, and General Intelligence.* Washington, DC: American Psychological Association, 2005.

Goldin-Meadow, Susan. "Talking and Thinking with Our Hands." *Current Directions in Psychological Science* 15 (2006): 34–39.

Goldman-Rakic, P. S. "Cellular Basis of Working Memory." *Neuron* 14 (March 1995): 477–85.

Gould, Stephen Jay *The Mismeasure of Man*, revised and expanded. New York: W. W. Norton & Company, 1996.

———. *Rock of Ages: Science and Religion in the Fullness of Life.* New York: Ballantine Publishing Group, 1999.

———. *The Structure of Evolutionary Theory.* Cambridge, MA: Harvard University Press, 2002.

Graffin, Gregory W., and William B. Provine. "Evolution, Religion, and Free Will." *American Scientist* 95 (July–August 2007): 294–97.

Greene, Joshua D., R. Brian Sommerville, Leigh E. Nystrom, John M. Darley, and Jonathon D. Cohen. "An fMRI Investigation of Emotional Engagement in Moral Judgment." *Science* 293 (September 2001): 2105–2108.

Greenwald, Anthony G., T. Andrew Poehlman, Eric Luis Uhlmann, and Jahzarin R. Banaji. "Understanding and Using the Implicit Association Test: III. Meta-Analysis of Predictive Value." *Journal of Personality and Social Psychology* 97 (2009): 17–41.

Griffin, Donald R. *Animal Thinking.* Cambridge, MA: Harvard University Press, 1984.

Haidt, Jonathon "Morality," *Perspectives on Psychological Science* 3 (2008): 65–72.

Hare, Brian. "From Nonhuman to Human Mind: What Changed and Why?" *Current Directions in Psychological Science* 16 (2007): 60–64.

Harris, Christine R. "A Review of Sex Differences in Sexual Jealousy, Including Self-Report Data, Psychophysiological Responses, Interpersonal Violence, and Morbid Jealousy." *Personality and Social Psychology Review* 7 (2003): 102–28.

———. "The Evolution of Jealousy." *American Scientist* 92 (2004): 62–71.

Hill, Robert Sean, and Christopher W. Walsh. "Molecular Insights into Human Brain Evolution." *Nature* 437 (September 1, 2005): 64–67.

Holloway, Ralph, "Evolution of the Human Brain." In *Handbook of Human Symbolic Evolution.* Edited by Andrew Lock and Charles R. Peters, Oxford: Clarendon Press, 1996, pp. 74–125.

Howe, Mark L. "Memories from the Cradle." *Current Directions in Psychological Science* 12 (2003): 62–65.

Huttenlocher, Peter R., and Arun S. Dabholkar, "Regional Differences in Synaptogenesis in Human Cerebral Cortex." *The Journal of Comparative Neurology* 387 (1997): 167–78.

Hyman, Steven E., Robert C. Malenka, and Eric J. Nestler, "Neural Mechanisms of Addiction: The Role of Reward-Related Learning and Memory." *Annual Review of Neuroscience* 32 (2006): 565–98.

Iacoboni, Marco, Istvan Molnar-Szakacs, Vittorio Gallese, Giovanni Buccino, John C. Mazziotta. "Grasping the Intentions of Others with One's Own Mirror Neuron System." *PLoS Biology* 3 e79 (March 2005): 0529–35.

Inzlicht, Michael, Ian McGregor, Jacob B. Hirsh, and Kyle Nash. "Neural Markers of Religious Conviction." *Psychological Science* 20 (2009): 385–92.

Izard, Carl. "Emotion Theory and Research: Highlights, Unanswered Questions, and Emerging Issues." *Annual Review of Psychology* 60 (2009): 1–25.

Jackson, Maggie. *Distracted: The Erosion of Attention and the Coming Dark Age.* Amherst, NY: Prometheus Books, 2008.

James, William. *Varieties of Religious Experience: A Study in Human Nature.* New York: Collier Books, 1961.

Johnson, Marcia K. "Memory and Reality." *American Psychologist* 61 (November 2006): 760–70.

Jonas, Eva, and Peter Fischer. "Terror Management and Religion: Evidence That Intrinsic Religiousness Mitigates Worldview Defense Following Mortality Salience." *Journal of Personality and Social Psychology* 91 (2006): 553–67.

Joorman, Jutta. "Cognitive Inhibition and Emotion Regulation in Depression." *Current Directions in Psychological Science* 19 (2010): 161–63.

Juttenlocher, Peter R., and Arun S. Dabholkar, "Regional Differences in Synaptogenesis in Human Cerebral Cortex," *The Journal of Comparative Neurology* 387 (1997): 167–78.

Kapogiannis, Dimitrios, Aron K. Barbey, Michael Su, Giovanna Zamboni, Frank Krueger, and Jordan Grafman. "Cognitive and Neural Foundations of Religious Belief." *Proceedings of the National Academy of Sciences* 106 (March 2009): 4876–81.

Kassin, Saul, Steven Fein, and Hazel Rose Markus. *Social Psychology*, 7th ed. Boston: Houghton Mifflin Company, 2008.

Kellogg, Ronald T. *Cognitive Psychology*, 2nd ed. Thousand Oaks, CA: Sage Publications, 1995.
———. "Training Writing Skills: A Cognitive Developmental Perspective." *Journal of Writing Research* 1 (2008): 1–26.

Kemps, Eva, and Marika Tiggemann. "A Cognitive Experimental Approach to Understanding and Reducing Food Cravings." *Current Directions in Psychological Science* 19 (2010): 86–90.

Klein, Richard G. *The Human Career: Human Biological and Cultural Origins*, 2nd ed. Chicago: University of Chicago Press, 1999.

Klein, Richard G., and Blake Edgar. *The Dawn of Human Culture.* New York: Nevraumont Publishing Company, 2002.

Klinger, Eric. *Daydreaming: Using Waking Fantasy and Imagery for Self-knowledge and Creativity.* Los Angeles: Jeremy P. Tarcher, 1990.

Koenigs, Michael, Liane Young, Ralph Adolphs, Daniel Tranel, Fiery Cushman, Marc Hauser, and Antonio Damasio. "Damage to the Prefrontal Cortex Increases Utilitarian Moral Judgments." *Nature* 446 (2007): 908–11.

Koole, Sander L., Jeff Greenberg, and Tom Pyszczynski, "Introducing Science to the Psychology of the Soul: Experimental Existential Psychology," *Current Directions in Psychological Science* 15 (2006): 212–16.

Koop, George F., and Michel Le Moal. "Addiction and the Brain Antireward System." *Annual Review of Psychology* 59 (2008): 29–53.

Krull, Douglas S., Michelle Hui-Min Loy, Jennifer Lin, Ching-Fu Wang, Suhong Chen, and Xudong Zhao. "The Fundamental Attribution Error: Correspondence Bias in Individualist and Collectivist Cultures." *Personality and Social Psychology Bulletin* 25 (1999): 1208–19.

Lai, Cecilia S. L., Dianne Gerrelli, Anthony P. Monaco, Simon E. Fisher, and Andrew J. Copp. "FOXP2 Expression during Brain Development Coincides with Adult Sites of Pathology in a Severe Speech and Language Disorder." *Brain* 126 (2003): 2455–62.

Landhuis, Carl Erik, Richie Poulton, David Welch, and Robert John Hancox. "Does Television Viewing Lead to Attention Problems in Adolescence? Results from a Prospective Longitudinal Study." *Pediatrics* 120 (2007): 532–37.

Lazar, Sara W., George Bush, Randy L. Gollup, Gregory L. Frichionne, Gurucharan Khalsa, and Herbert Benson. "Functional Brain Mapping of the Relaxation Response and Meditation." *NeuroReport* 11, no. 7 (2000): 1581–85.

Lazarus, Richard S. "Thoughts on the Relation between Emotion and Cognition." *American Psychologist* 37 (1982): 1019–24.

LeDoux, Joseph E. "Emotion Circuits in the Brain." *Annual Review of Neuroscience* 23 (2000): 155–84.

———. "Emotion, Memory, and the Brain." *Scientific American* 12, no. 1 (2002): 62–71.

Lehman, Darrin R., Chi-yie Chiu, and Mark Schaller. "Psychology and Culture." *Annual Review of Psychology* 55 (2004): 689–714.

Lewin, Roger. "Mitochondrial Eve: The Biochemical Route to Human Origins." *Mosaic* 22, no. 3 (1991): 46–63.

Li, Wen-Hsiung, and Matthew A. Saunders. "The Chimpanzee and Us," *Nature* 437 (September 1, 2005): 50–51.

Liberman, Phillip. *Eve spoke: Human Language and Human Evolution*. New York: W. W. Norton & Company, 1998.

Lieberman, Matthew D. "Social Cognitive Neuroscience: A Review of Core Processes," *Annual Review of Psychology* 58 (2007): 259–89.

Lindsay, D. Stephen, Lisa Hagen, J. Don Read, Kimberly A. Wade, and Maryanne Garry. "True Photographs and False Memories." *Psychological Science* 15 (2004): 149–54.

Lorber, Michael F. "Psychophysiology of Aggression, Psychopathy, and Conduct Problems: A Meta-Analysis." *Psychological Bulletin* 130 (2004): 531–52.

Loftus, Elizabeth F., and Katherine Ketcham. *The Myth of Repressed Memory*. New York: St. Martin's Press, 1994.

Lovejoy, C. Owen. "Reexamining Human Origins in Light of *Ardipithecus ramidus*." *Science* 326 (October 2, 2009): 74e1–e8, doi:10.1126/science.1175834.

Lutz, Antoine, Heleen A. Slagter. John D. Dunne, and Richard J. Davidson. "Attention Regulation and Monitoring in Meditation." *Trends in Cognitive Science* 12, no. 4 (2008): 163–69.

MacLean, Paul D. "On the Evolution of the Three Mentalities of the Brain." In *Origins of Human Aggression: Dynamics and Etiology*. Edited by Gerard G. Neuman. New York: Human Sciences Press, 1987, pp. 29–41.

Marche, Stephen. "The Book That Contains All Books." *Wall Street Journal* (October 17–18, 2009), p. W9.

Mark Mazower. *Hitler's Empire: How the Nazis Ruled Europe*. New York: Penguin Press, 2008.

Markus, Helen, and Elissa Wurf. "The Dynamic Self-Concept: A Social and Psychological Perspective." *Annual Review of Psychology* 38 (1987): 299–337.

Mazzoni, Giuliana A. L., Elizabeth L. Loftus, and Irving Kirsch. "Changing Beliefs about Implausible Autobiographical Events: A Little Implausibility Goes a Long Way." *Journal of Experimental Psychology: Applied* 7 (2001): 51–59.

McBrearty, Sally and Alison S. Brooks. "The Revolution that Wasn't: A New Interpretation of the Origin of Modern Human Behavior." *Journal of Human Evolution* 39 (2002): 453–563.

McCullough, Michael E., and Brian L. B. Willoughby. "Religion, Self-Regulation, and Self-Control: Associations, Explanations, and Implications." *Psychological Bulletin* 135 (2009): 69–93.

Meltzoff, Andrew N., Patricia K. Kuhl, Javier Movellan, and Terrence J. Sejnowski. "Foundations for a New Science of Learning." *Science* 325 (July 2009): 284–88.

Micklethwait, John, and Adrian Wooldridge. *God is Back: How the Global Revival of Faith Is Changing the World*. New York: Penguin Press, 2009.

Milner, Brenda. "Amnesia following Operations on the Temporal Lobes." In *Amnesia*. Edited by C. W. M. Whitty and O. L Zangwill. London: Butterworth, 1966, pp. 109–33.

Mischel, Walter, Y. Shoda, and P. K. Peake. "The Nature of Adolescent Competencies Predicted by Preschool Delay of Gratification." *Journal of Personality and Social Psychology* 54 (1988): 687–96.

Mithen, Steven. *The Prehistory of the Mind: The Cognitive Origins of Art, Religion, and Science*. London: Thames and Hudson, 1996.

Miyake, Akira, and Naomi P. Friedman. "The Nature and Organization of Individual Differences in Executive Functions: Four General Conclusions." *Current Directions in Psychological Science* 21 (2012): 8–14.

Moscovitch, Morris, Gordon Winocur, and Marlene Behrmann. "What Is Special about Face Recognition? Nineteen Experiments on a Person with Visual Object Agnosia and Dyslexia but Normal Face Recognition." *Journal of Cognitive Neuroscience* 9 (1997): 555–604.

Mundy, Peter, and Lisa Newell. "Attention, Joint Attention, and Social Cognition." *Current Directions in Psychological Science* 16 (2007): 269–74.

Neisser, Ulric. "John Dean's Memory." In *Memory Observed: Remembering in Natural Contexts*, 2nd ed. Edited by Ulric Neisser and Ira E. Hyman Jr. New York: Worth Publishers, 2000, pp. 263–86.

Newberg, Andrew, Eugene G. d'Aquili, and Vince Rause. *Why God Won't Go Away: Brain Science and Biology of Believing*. New York: Ballantine Books, 2001.

Nielsen, Michael. *Reinventing Discovery*. Princeton, NJ: Princeton University Press, 2012.

Nolen-Hoeksema, Susan, Blair E. Wisco, and Sonja Lyubomirsky. "Rethinking Rumination." *Perspectives on Psychological Science* 3 (2008): 400–24.

Ogden, Cynthia L., Margaret D. Carroll, Brian K. Kit, and Katherine M. Flegal. "Prevalence of Obesity in the United States, 2009–2010." NCHS Data Brief No. 82. http://www.cdc.gov/nchs/data/databriefs/db82.pdf (accessed January 2012): 1–7.

Oltmanns, Thomas F., and Robert E. Emery. *Abnormal Psychology*, 4th ed. Upper Saddle River, NJ: Pearson Education, 2004.

Olson, David R. *The World on Paper: The Conceptual and Cognitive Implications of Writing and Reading*. New York: Cambridge University Press, 1994.

Olson, Steve. *Mapping Human History: Genes, Race, and Our Common Origins*. Boston: Houghton, Mifflin Company, 2002.

Ong, Walter. *Orality and Literacy: The Technologizing of the Word*. London: Methuen, 1982.

Parrot, André. *The Tower of Babel*. New York: Philosophical Library, 1955.

Phelps, Elizabeth A. "Emotion and Cognition: Insights from the Study of the Human Amygdala." *Annual Review of Psychology* (2006): 27–53.

Poldrack, Russell A., and Anthony D. Wagner, "What Can Neuroimaging Tell Us about the Mind: Insights from Prefrontal Cortex." *Current Directions in Psychological Science* 13 (2004): 177–80.

Poldrack, Russell A., and Anthony D. Wagner. "What Can Neuroimaging Tell Us about the Mind: Insights from Prefrontal Cortex." *Current Directions in Psychological Science* 13 (2004): 177–81.

Ponting, Chris P., and Gerton Lunter. "Human Brain Gene Wins Genome Race." *Nature* 443 (September 2006): 149–50.

Posner, Michael I., and Mary K. Rothbart, *Educating the Human Brain*. Washington, DC: American Psychological Association, 2007.

Price, Donald D., Damien G. Finniss, and Fabrizio Benedetti. "A Comprehensive Review of the Placebo Effect: Recent Advances and Current Thought." *Annual Review of Psychology* 59 (2008): 565–90.

Proust, Melanie, Rebecca Bellone, Norbert Benecke, Edson Sandoval-Castellanos, Michael Cieslak, Tatyana Kuznetsova, Arturo Morales-Muniz, et al. "Genotypes of Predomestic Horses Match Phenotypes Painted in Paleolithic Works of Cave Art." *Proceedings of the National Academy of Sciences* 108 (2011): 18626–30.

Raven, Peter H., and George B. Johnson. *Biology*, 5th ed. Boston: McGraw-Hill, 1999.

Reader, Simon M., and Kevin N. Laland "Social Intelligence, Innovation, and Enhanced Brain Size in Primates." *Proceedings of the National Academy of Sciences* 99 (April 2, 2002): 4436–41.

Ridley, Matt. "Humans: Why They Triumphed." *Wall Street Journal*, May 22–23, 2010, pp. W1–W2.

Robinson, Terry E., and Kent C. Berridge. "Addiction." *Annual Review of Psychology* 54 (2003): 25–53.

Roser, Matthew E., Jonathan A. Fugelsang, Kevin N. Dunbar, Paul M. Corballis, and Michael E. Gazzaniga. "Dissociating Processes Supporting Causal Perception and Causal Inference in the Brain." *Neuropsychology* 19 (2005): 591–602.

Roubini, Nouriel, and Stephen Mihm, *Crisis Economics: A Crash Course in the Future of Finance.* New York: Penguin Press, 2010.

Rubin, David. C. "The Basic-Systems Model of Episodic Memory." *Perspectives on Psychological Science* 1 (2006): 277–311.

———. *Memory in Oral Traditions: The Cognitive Psychology of Epic, Ballads, and Counting-out Rhymes.* New York: Oxford University Press, 1995.

Ruhlen, Merritt. *The Origin of Language: Tracing the Evolution of the Mother Tongue.* New York: John Wiley and Sons, 1994.

Savage-Rumbaugh, E. Sue, and Duane M. Rumbaugh, "The Emergence of Language." In *Tools, Language and Cognition in Human Evolution.* Edited by Kathleen R. Gibson and Tim Ingold. Cambridge: Cambridge University Press, 1993, pp. 86–108.

Schmandt-Bessarat, Deborah, "Tokens as Precursors of Writing." In *Writing: A Mosaic of New Perspectives.* Edited by Elena L. Grigorenko, Elisa Mambrino, David D. Preiss. New York: Psychology Press, 2012, pp. 3–10.

Schwartz, Bennett L. "Do Nonhuman Primates Have Episodic Memory?" In *The Missing Link in Cognition.* Edited by Herbert S. Terrace and Janet Metcalfe. Oxford, Oxford University Press, 2005, pp. 225–41.

Schwartz, Seth J., Jennifer B. Unger, Byron L. Zamboanga, and José Szapocznik. "Rethinking the Concept of Acculturation: Implications for Theory and Research." *American Psychologist* 65 (2010): 237–51.

Sedikides, Constantine, Tim Wildschut, Jamie Arndt, and Clay Routledge. "Nostalgia: Past, Present, and Future." *Current Directions in Psychological Science* 17 (2008): 304–307.

Semendeferi, K., A. Lu, N. Schenker, and H. Damasio. "Humans and Great Apes Share a Large Frontal Cortex." *Nature Neuroscience* 5 (March 2002): 272–76.

Seybold, Kevin S. and Peter C. Hill. "The Role of Religion and Spirituality in Mental and Physical Health." *Current Directions in Psychological Science* 10 (2001): 21–24.

Singer, Dorothy G., and Jerome R. Singer, *Imagination and Play in the Electronic Age.* Cambridge, MA: Harvard University Press, 2005.

Smith, Edward E., and John Jonides. "Working Memory: A View from Neuroimaging." *Cognitive Psychology* 33 (1997): 5–42.

Solomon, Sheldon, Jeff Greenberg, and Tom Pyszczynski, "Pride and Prejudice: Fear of Death and Social Behavior." *Current Directions in Psychological Science* 9 (2000): 200–204.

Sowell, Elizabeth R., Paul M. Thompson, Colin J. Holmes, Terry L. Jernigan, and Arthur W. Toga. "In Vivo Evidence for Post-Adolescent Brain Maturation in Frontal and Striatal Regions." *Nature Neuroscience* 2 (October 1999): 859–61.

Spanos, Nicholas P. *Multiple Identities and False Memories: A Sociocognitive Perspective*. Washington, DC: American Psychological Association, 1996.

Szpunar, Karl K., Jason M. Watson, and Kathleen B. McDermott. "Neural Substrates of Envisioning the Future." *Proceedings of the National Academy of Science* 104 (January 2007): 642–47.

Steeves, Jennifer, Laurence Dricot, Herbert C. Goltz, Bettina Sorger, Judith Peters, A. David Milner, Melvyn A. Goodale, et al. "Abnormal Face Identity Coding in the Middle Fusiform Gyrus of Two Brain-Damaged Prosopagnosic Patients." *Neuropsychologia* 47 (2009): 2584–92.

Steinberg, Laurence "Risk Taking in Adolescence: New Perspectives from Brain and Behavioral Science" *Current Directions in Psychological Science* 16 (2007): 56–59.

Swing, Edward L., David A. Gentile, Craig A. Anderson, and David A. Walsh. "Television and Video Game Exposure and the Development of Attention Problems." *Pediatrics* 126 (2010): 214–21.

Suddendorf, Thomas, and Michael C. Corballis. "The Evolution of Foresight: What Is Mental Time Travel, and Is It Unique to Humans?" *Behavioral and Brain Sciences* 30 (June 2007): 299–313.

Tapscott, Don, and Anthony D. Williams. *Wikinomics: How Mass Collaboration Changes Everything*. New York: Portfolio, 2006.

Taylor, Jill Bolte. *My Stroke of Insight: A Brain Scientist's Personal Journey*. New York: Viking, 2006.

Taylor, Shelley E. *Positive Illusions: Creative Self-Deception and the Health Mind*. New York: Basic Books, 1989.

———. "Tend and Befriend: Biobehavioral Bases of Affiliation under Stress." *Current Directions in Psychological Science* 15 (2006): 273–77.

Taylor, Shelley E., and Jonathon D. Brown. "Illusion and Well-Being: A Social Psychological Perspective on Mental Health." *Psychological Bulletin* 103 (1988): 193–210.

Templeton, Alan R. "Human Races: A Genetic and Evolutionary Perspective." *American Anthropologist* 100 (1999): 632–50.

Thompson, Richard F. *The Brain: A Neuroscience Primer*, 3rd ed. New York: Worth Publishers, 2000.

Tippett, Krista. *Speaking of Faith*. New York: Viking Press, 2007.

Tomasello, Michael. *The Cultural Origins of Human Cognition*. Cambridge, MA: Harvard University Press, 1999.

Tulving, Endel. "Episodic Memory and Autonoesis: Uniquely Human?" In *The Missing Link in Cognition.* Edited by Herbert S. Terrace and Janet Metcalfe. Oxford, Oxford University Press, 2005, pp. 3–56.

———. "Episodic Memory: From Mind to Brain." *Annual Review of Psychology* 53 (2002): 1–25.

Turk, David J., Todd F. Heatherton, William M. Kelley, Margaret G. Funnell, Michael S. Gazzaniga, and C. Neil Macrae. "Mike or Me? Self-Recognition in a Split Brain Patient." *Nature Neuroscience* 9 (September 2002): 841–42.

Valkenburg, Patti M., and Joehen Peter. "Social Consequences of the Internet for Adolescents: A Decade of Research." *Current Directions in Psychological Science* 18 (2009): 1–5.

Vargas-Cooper, Natasha. "Hard Core: The New World of Porn Is Revealing Eternal Truths about Men and Women." *Atlantic*, January/February 2011, pp. 97–106.

Vygotsky, Lev. S. *Mind in Society: The Development of Higher Psychological Processes.* Edited by Michael Cole, Vera John-Steiner, Sylvia Scribner, Ellen Souberman. Cambridge, MA: Harvard University Press, 1978.

———. *Thought and Language.* Translated, newly revised, and edited by Alex Kozulin. Cambridge, MA: MIT Press, 1986.

Waller, Niels G., Brian A. Kojetin, Thomas J. Bouchard Jr., David T. Lykken, and Auke Tellegen, "Genetic and Environmental Influences on Religious Interests, Attitudes, and Values: A Study of Twins Reared Apart and Reared Together," *Psychological Science* 1 (1990): 138–42.

Wexler, Bruce E. *Brain and Culture.* Cambridge, MA: MIT Press, 2006.

Wheatley, Thalia, and Jonathan Haidt. "Hypnotic Disgust Makes Moral Judgments More Severe." *Psychological Science* 16 (2005): 780–84.

Wheeler, Mary E., and Susan T. Fiske. "Controlling Racial Prejudice: Social-Cognitive Goals Affect Amygdala and Stereotype Activation." *Psychological Science* 16 (2005): 56–60.

White, Randall. *Prehistoric Art: The Symbolic Journey of Humankind.* New York: Harry N. Abrams, 2003.

———. "On the Evolution of Human Socio-Cultural Patterns." In *Handbook of Human Symbolic Evolution.* Edited by Andrew Lock and Charles R. Peters. Oxford: Clarendon Press, 1996, pp. 239–62.

Whiten, Andrew. "The Second Inheritance System of Chimpanzees and Humans." *Nature* 437 (September 2005): 52–55.

Whitson, Jennifer A., and Adam D. Galinsky. "Lacking Control Increases Illusory Pattern Perception." *Science* 322 (October 2008): 115–17.

Wolford, George, Michael B. Miller, and Michael Gazzaniga. "The Left Hemisphere's Role in Hypothesis Formation." *Journal of Neuroscience* 20 (2000): RC64 (1–4).

Woodard, Emory H, IV. *Media in the Home 2000: The Fifth Annual Survey of Parents and Children.* Philadelphia, PA: Annenberg Public Policy Center of the University of Pennsylvania, Survey Series No. 7.

Wynn, Thomas, and Frederick L. Coolidge. "A Stone-Age Meeting of Minds." *American Scientist* 96 (January–February 2008): 44–51.

Wu, Xiaohong, Chi Zhang, Paul Goldberg, David Cohen, Yan Pan, Trina Arpin, and Ofer Bar-Yosef. "Early Pottery at 20,000 Years Ago in Xianrendong Cave, China." *Science* 336 (June 2012): 1696–1700.

Yoo, Hee Jeong, Soo Churl Cho, Jihyun Ha, Sook Kyung Yune, Seog Ju Kim, Jaeuk Hwang, Ain Chung, Young Hoon Sung, and In Kyoon Lyoo. "Attention Deficit Hyperactivity Symptoms and Internet Addiction." *Psychiatry and Clinical Neurosciences* 58 (2004): 487–94.

Zuberbühler, Klaus. "The Phylogenetic Roots of Language: Evidence from Primate Communication and Cognition." *Current Directions in Psychological Science* 14 (2005): 126–30.

INDEX